Careers in Forest, Wildlife, Fisheries, and Range Resources

Careers in Forest, Wildlife, Fisheries, and Range Resources

Ron Boldenow
Central Oregon Community College

WAVELAND
PRESS, INC.
Long Grove, Illinois

For information about this book, contact:
Waveland Press, Inc.
4180 IL Route 83, Suite 101
Long Grove, IL 60047-9580
(847) 634-0081
info@waveland.com
www.waveland.com

Copyright © 2019 by Waveland Press, Inc.

10-digit ISBN 1-4786-3625-4
13-digit ISBN 978-1-4786-3625-0

All rights reserved. No part of this book may be reproduced, stored in a retrieval system, or transmitted in any form or by any means without permission in writing from the publisher.

Printed in the United States of America

7 6 5 4 3 2 1

This book is dedicated to my wife, Lauri,
and to my biology and forestry professors at
Calvin College, Humboldt State University, and
the University of California, Berkeley—most notably
John Beebe, Bill Bigg, Joe McBride, and John Helms.
Thank you.

Contents

Preface   xi

1 Passion, Path, Plan, Career — 1

What Are Natural Resources?	3
Public Trust Doctrine and Social License	5
Career, Profession, and Occupation	6
Topics and Their Organization in This Book	8

WORKS CITED   9

2 Professions — 10

Professional Science-Based Work   12

Natural Resources Management and Biological Sciences (400 Group)   13

- The Forestry Profession   14
- The Range Manager Profession   18
- The Fisheries Manager Profession   21
- The Wildlife Manager Profession   24

Related Professions   30

- General Natural Resources Management and Biological Sciences (401 Series)   30
- Professional Work Compared to Technical Jobs   31
- Range Technician (455 Series)   32

  Forest Technician (462 Series) 33
  Park Ranger Profession (25 Series) 34
  Wildland Fire 36
Federal Tours of Duty and Position Types 39
Federal General Schedule Education
  Requirements and Salaries 40
Summary 41
STUDENT EXERCISES 41 WORKS CITED 42 ADDITIONAL RESOURCES 43

3   History   44

Developing the Idea of Conservation,
  Civil War to 1890 46
Conservation and Wise Use, 1891 to 1930 49
Management Period, 1930 to 1960 53
Concern about Quality of Life,
  Late 1950s to 1970 55
Global Concerns, 1970s to Present Day 57
Summary 59
STUDENT EXERCISES 60 WORKS CITED 60 ADDITIONAL RESOURCES 61

4   Agencies   62

United States Federal Government 64
  Department of Interior 66
  Department of Agriculture 68
  Department of Defense 69
  Department of Commerce 70
  Department of Energy (DOE) 70
  Environmental Protection Agency (EPA) 71
  Quasi-Government Agencies 71

Contents

Tribal Governments and Alaska Native Corporations   71
State and Local Governments   71
    State Governments   71
    Local Governments   72
Other Employers of Natural Resources Professionals   73
    Nongovernment Organizations   73
    Educational Organizations   73
    Private Landowners and Managers   73
Summary   75
STUDENT EXERCISES 75   WORKS CITED 75

5 EDUCATION   77

Academic Definitions   78
Natural Resources Majors   81
Components of an Academic Curriculum
    for a Bachelor's Degree in Natural Resources   82
Accreditation   87
Professional Certification   87
The Future of Natural Resources Education   88
Continuing Education   92
Top Ten Tips for Potential Students
    in the Natural Resources   92
Summary   95
STUDENT EXERCISES 96   WORKS CITED 96   ADDITIONAL RESOURCES 97

6 PRACTICAL MATTERS   98

Preparing for a Career   99
    Gaining Experience   99
    Applying for Jobs   99

  Interviewing 100
  Providing References 101
  Tips from a Career Guide 102
Human Dimension Skills   **104**
  Communication 104
  Ethics 105
  Basic and Professional Etiquette 105
  Working in Groups 106
  Records and Accounting 106
Natural Resources Field Skills   **108**
  Local Ecosystem Characteristics 108
  Basic Land and Water Navigation 109
  Outdoor Safety and Efficiency 109
Summary   **115**
STUDENT EXERCISES 115 WORKS CITED 115

Appendix A: Accreditation Standards 117
Appendix B: Certification Requirements 129
Index 141

# Preface

The concept for this book grew out of a discussion between Don Rosso of Waveland Press and myself at the national convention of the Society of American Foresters in Madison, Wisconsin, in 2016. My suggestion to Don was to add a chapter about professional education and development to an introductory forestry textbook. Instead, the suggestion evolved into an idea for a short book that could be used as a supplemental text in an introductory course in forestry, wildlife, or fisheries. Along the way, the discipline of range management was added to the book. This guide can also be used in a stand-alone short course to orient students to natural resources professions. High school guidance counselors should also find this to be a valuable tool.

My teaching career has spanned 19 years of teaching, advising, and recruiting at Central Oregon Community College, as well as part-time teaching for another community college and two universities. I am a generalist, having taught more than 20 different courses in forestry and natural resources. In addition, I have been active in the Society of American Foresters, serving as a local chapter chair, chair of the Oregon Society of American Foresters, and four years on the national Committee on Forest Technology School Accreditation. I have also served on forestry advisory committees for two universities and participated in the North American Summit on Forest Science Education held at the University of California, Berkeley, in May 2014. These experiences have provided insight into both the expectations of students seeking to prepare for a career in natural resources and the expectations of employers wishing to hire students who are well prepared to begin their careers.

The underlying assumption of this book is that students who have a greater understanding of what is expected in college and within the profession will be better equipped to succeed in school and on the job. The purpose of this book is to provide both educational and professional context to students wishing to pursue a natural resources career.

In my experience, many such students are the first in their families to either attend college or to work in the field of natural resources. Many of these students do well and achieve success, but others do not. This book is an attempt to orient students to the various traditional natural resources disciplines of forestry, wildlife management, fisheries management, and range management and to prepare them to be successful in their pursuit of an education and a career.

I wish to thank my wife, Lauri, and colleagues Rebecca Franklin, Paula Simone, and Bret Michalski; each offered suggestions on content and reviewed portions of the text. A special thank you to Waveland Press editor Laurie Prossnitz for her insightful questions and diligent help in editing the text.

Passion, Path, Plan, Career

Chapter Outline
What Are Natural Resources?
Public Trust Doctrine and Social License
Career, Profession, and Occupation
Topics and Their Organization in This Book

"You will succeed at whatever you have a passion for doing." That is the advice often given to college students seeking to discover a career path. It is good and true advice. Many students have a strong passion for the outdoors—for trees, forests, wildlife, hunting, fishing, backpacking, climbing, boating, or other types of wildland recreation. Hopefully they also have a passion for science, for continuous learning, and for service to society. They know they wish to pursue a career in natural resources, but know little of the possible professions. Other students know a bit more about their options, but are not sure how to chart a course to their desired career. If you have read this far, then you or someone you know is interested in a career in natural resources. Hopefully this interest will develop into a passion.

Natural resource careers entail working outdoors, indoors, alone, in teams, and with the public, but always for the benefit of the resource and society. They require a wide range of knowledge and skills as well as a commitment to lifelong learning. There is no single route into a natural resources career. There are, however, paths that are more efficient with regard to the investment of effort, time, and money to gain education and experience. The goal of this book is to help you choose an efficient path to a successful career in forestry, wildlife management, fisheries management, or range management.

Sacramento River Bend, a Bureau of Land Management (BLM) Outstanding Natural Area in northern California, contains riparian and oak savannah environments. Not all natural resources professionals work in such scenic settings, but many do. (Photo by Bob Wick, BLM, Flickr)

This book is geared to students entering a community college or university, but should prove useful to high school students and even graduate students contemplating a career in natural resources. The purpose is to give information, advice, and guidance on pursuing a career in forestry, wildlife management, fisheries management, or range management—the four traditional natural resource professions that are the primary focus of this book. Many potential students, and even those well into their studies, don't know how to channel their interests to build a career. They know that they want to "work in the woods" or "on the water." Students say they want to be a "forest ranger" or a "game warden" without fully understanding what those careers entail. Building a successful career takes time, effort, and support from family, friends, and employers. This book will help you become more knowledgeable and articulate about natural resource career possibilities. As with any career path, there may be pitfalls and drawbacks. But you will succeed at whatever you have a passion for doing. Attend a meeting of foresters, wildlife managers, fisheries managers, or range managers and you will observe not only their passion for the resources they study, protect, and manage, but also enthusiasm for their profession.

Before going further a few concepts need to be explained and some definitions provided. The rest of this chapter will include a broad overview of natural resources, natural resource professions, and the topics that will be covered in succeeding chapters.

WHAT ARE NATURAL RESOURCES?

Natural resources are naturally occurring substances that are used and given value by humans. These resources can be **renewable**, such as vegetation that grows back after being harvested. For example, the grass grazed by cows is an example of a resource that is extracted from the environment but may be renewed if managed properly. **Nonrenewable** resources are those that are not replaced after extraction. The mining of coal and iron ore are good examples of nonrenewable natural resources. Some resources are nonextractive. An example would be the use of a forest setting for recreation. Bird watching is one example of a nonextractive use of a forest resource. This does not mean that the activity does not influence the resource; it simply means the resource is not extracted from its environment. The behavior of the birds may change with the presence of people, but the bird remains in its environment.

The US Forest Service is an agency within the US Department of Agriculture, and manages the second largest land area in the United States. The agency's mission is to sustain the health, diversity, and productivity of the nation's forests and grasslands to meet the needs of present and future generations. Much of its work in carrying out this mission involves the five common renewable natural resources: wood, water, wildlife, range, and recreation (jokingly called the five W's with wange and wecreation). The importance of these five resources cannot be overstated. It is equally important to have qualified natural resource managers to oversee them. Such resources exist not only on the lands of the national forests, but on all wildlands, public and private. Much

Some tasks for natural resources professionals require office work, like preparing management plans. (Photo by BLM Alaska, Flickr)

of what foresters, wildlife managers, fisheries managers, and range managers do on the job involves managing the wood, water, wildlife (including fish), range, and recreational opportunities on a particular land area, either public or private.

The natural resources of the United States are extensive but also heavily utilized. Currently the United States has 5 percent of the world's population, 7 percent of the world's land area, 8 percent of its forestland, 6 percent of its woody biomass, and 10 percent of its timber industry, and yet Americans consume 28 percent of the world's industrial wood products (US Forest Service 2014). Additionally, forests are the source for over half of the water used in the United States, and the very idea of wild areas, wilderness, and national parks are an integral part of the American culture.

According to the *Dictionary of Forestry*, **forests** are ecosystems with dense and extensive tree cover; **woodlands** may mean a forest area or, more specifically, an area with small, short trees with open, often grassy, areas between the trees; and **grasslands** are dominated by grass and grass-like plants (Helms 1998). In 2012, the United States had 766 million acres of **forestland**, about 33 percent of our total land area. This is about the same area that was in forestland in 1920 (US Forest Service 2014). Of this forestland, 521 million acres are timberland used for the production of wood products, 74 million acres are in some type of protected preserve, and 172 million acres are in some other type of forest. There is some overlap in these categories, with some land classified in two categories. Public forestlands dominate in the western states, private forestlands dominate in the eastern states, and private, industrial timberland is concentrated in the southeastern states, the Pacific Northwest, upper Great Lakes states, and in New England. In addition, some 53 million acres of the United States is in woodlands—areas that do not have as dense a tree cover as forests. Urban forests are also important. About 3.1 percent of the land area of the United States is classified as urban area and populated by 220 million people, or about 80 percent of the country's population.

The rangeland area of the United States is about 36 percent of our total land area or about 770 million acres (US Forest Service n.d.). **Rangeland** is land suitable for grazing livestock but is not suitable for plowing, so intense agriculture is not possible on this land. Over half of the rangeland is in private ownership, and about 43 percent is owned by the federal government. States and local governments own the rest. Much of this rangeland is interspersed or adjacent to forestland.

The surface area of the inland and coastal waters of the United States is about 169 million acres (US Census Bureau 2012). Within this area live numerous varieties of fish and other aquatic animals.

Wildlife inhabits all of these forest, grassland, and aquatic ecosystems. There have been successes and losses in the preservation of game and nongame wildlife in the last century. Big game populations are generally increasing as is the population of waterfowl (Flather et al. 2013).

Recreation is an important component of natural resource use. The greatest recreational use of the forests in the United States is walking for pleasure, with an estimated 8.5 billion recreation activity days in 2015, of which 45 percent occurred in urban forests (US Forest Service 2014). A **recreation activity day** is the unit of measure for recreation and is the amount of activity completed by one person in one day. That is an astounding number—more than 3.5 billion recreation activity days of people walking for pleasure in urban forests in a year. The other top-ranked forest activities are also nonconsumptive; viewing scenery, photographing or viewing birds, and viewing other wildlife are the three next most frequent reasons for a recreation visit to a forest. Additionally, the National Park Service reported nearly 331 million visits to its units in 2017 (National Park Service n.d.).

Because our natural resources are heavily utilized, they need to be properly managed. The responsibility of this management falls to the foresters, wildlife managers, fisheries managers, and range managers who are professionally trained with the knowledge and skills to manage their resource in the context of our society's needs and desires. The requirement that natural resources professionals need to understand the social context of their actions will be a recurring theme in this book. There is a saying among some in the industry that natural resource management is not rocket science—it is much more complex! Much, but not all, of that complexity occurs because natural resource managers must work within a social arena.

Public Trust Doctrine and Social License

Public trust doctrine is the concept that the sovereign holds some resources, including some natural resources, in public trust regardless of private property rights. The doctrine is based in ancient law and was incorporated into US law through England's influence on our culture and legal system. The best example is the regulation of game species by individual states and territories. A deer running across, or bedded down, on private property is not owned by the property owner. The deer is held in public trust by the state for all to enjoy. The federal government also holds the wildlife on its land in public trust and can supersede state law to protect that wildlife. Natural resources are managed in the context of public trust both on government managed land and on private land, and the resources of wildlife, water, and air remain in the public trust.

A **social license** is society's permission to do something. There is a social license for many activities, including the practice of grazing cattle, suppressing wild fires, harvesting timber, prescribed burning, hunting, and catching fish, but the social license for these activities is constantly changing. Society has directed that forestry should no longer be practiced in the same manner as it was in the 1950s, and comprehensive forest-practice laws have been enacted in many states. Another example of a social license change affecting natural resources is the trapping of fur-bearing animals. Even though fur bearers may be legally trapped and are a renewable resource, the wearing of fur is no longer socially acceptable in much of society. Using furs to make apparel thus has a declining social license. This declining social license has depressed the fur market dramatically, and trapping is not as profitable as it once was. In natural resource management, social license and the social context are closely intertwined in decision making.

CAREER, PROFESSION, AND OCCUPATION

A **career** is defined as "an occupation undertaken for a significant period of a person's life and with opportunities for progress" (Google n.d.). In other words, a career is ever changing. A **profession** is defined by the same source as "a paid occupation, especially one that involves prolonged training and a formal qualification or the body of people engaged in a particular profession." A synonym for both terms is a **calling,** which is defined by the dictionary as "a strong urge toward a particular way of life or career; a vocation." Further characteristics of a profession are that it is a full-time occupation, requires special training in a college setting, and has its own professional associations, a code of ethics, and licensing requirements.

Forestry, wildlife management, fisheries management, and range management are science-based professions and have been taught as separate disciplines for decades. They all have a similar foundation in biology and ecology, but each has developed distinct skill sets and spheres of influence. They have separate professional societies with ethical standards, and have different forms of credentialing and state licensing requirements. Despite these differences, the people in these professions generally perform similar processes—that is, measuring and monitoring the resource and then determining how to best manage it through harvest or protection. According to Robert Froese of Michigan Technological University, "if you can't measure it, you cannot manage it!" But sometimes just measuring the resource is a costly and difficult task; for example, sampling fish that are difficult to locate and catch. Another similarity between the professions is their need to analyze data, interpret the results, and then apply the interpretation to manage the resource. This analytical component is similar across the four professions, but each applies advanced knowledge and skill sets that are distinct.

Forestry is defined in the *Dictionary of Forestry* (Helms 1998) as "the profession embracing the science, art, and practice of creating, managing, using, and conserving forest and associated resources for human benefit in a sustainable manner to meet desired goals, needs, and values." The dictionary notes that "the broad field of forestry consists of those biological, quantitative, managerial, and social sciences that are applied to forest management and conservation."

Wildlife management and **fisheries management** are defined as the professions embracing the science, art, and practice of creating, managing, using, and conserving wildlife or fish resources and their habitat for both the animals' and human benefit. To streamline the vocabulary within this book, the term *wildlife management* will be used to mean both wildlife management and wildlife biology. To stay consistent, the term *fisheries management* will be used to mean both fisheries management and fisheries science.

Range management is the profession dealing with the use of rangelands and range resources for a variety of purposes. These purposes

Sometimes natural resource fieldwork is just fun. (Photo by Deanna Younger, BLM New Mexico Field Office, Flickr)

"If you can't measure it, you can't manage it," says Robert Froese of Michigan Technological University. Here, a fish biologist with the National Park Service (NPS) measures fish as part of a project on Quartz Lake in Glacier National Park. (Photo by James Greig, NPS, Flickr)

include watersheds, wildlife habitat, grazing by livestock, recreation, and aesthetics. Again, this is practiced for human benefit.

Topics and Their Organization in This Book

The chapters of this guide can stand alone or be used in combination. Most include a list of works cited that allow further investigation into topics of interest. Some chapters also feature student exercises.

Chapter 2 provides extensive descriptions of the four traditional natural resource professions that are the focus of this text: forestry, wildlife management, fisheries management, and range management. Descriptions of other natural resource professions will be provided as well for comparison and because of their popularity.

Chapter 3 provides historical context and briefly describes the history of the conservation movement, the development of natural resource policy, the science of ecology, and the evolution of the traditional four natural resource professions.

Chapter 4 discusses the primary employers of natural resources professionals, while Chapter 5 summarizes the requirements of college programs in forestry, wildlife management, fisheries management, and range management. Other natural resource programs will be included to allow comparisons. Finally, Chapter 6 offers practical advice for those hoping to develop careers in natural resources.

WORKS CITED

Flather, C. H., Knowles, M. S., Jones, M. F., and Schilli, C. 2013. *Wildlife Population and Harvest Trends in the United States: A Technical Document Supporting the Forest Service 2010 RPA Assessment*. Gen. Tech. Rep. RMRS-GTR-296. Fort Collins, CO: US Department of Agriculture, Forest Service, Rocky Mountain Research Station.

Google. n.d. *Google Dictionary*. Accessed September 18, 2017. https://www.google.com.

Helms, J. A. 1998. *The Dictionary of Forestry*. Bethesda, MD: Society of American Foresters.

National Park Service. n.d. *Social Science: Annual Visitation Highlights*. Accessed April 19, 2018. https://www.nps.gov/subjects/socialscience/annual-visitation-highlights.htm.

US Census Bureau. 2012. *Geography, State Area Measurements and Internal Point Coordinates*. Last revised December 5, 2012. https://www.census.gov/geo/reference/state-area.html.

US Forest Service. n.d. *About Rangeland Management*. Accessed January 26, 2018. https://www.fs.fed.us/rangeland-management/aboutus/index.shtml.

US Forest Service. 2014. *US Forest Resource Facts and Historical Trends*. FS-1035. Washington, DC: US Department of Agriculture.

2

Professions

Chapter Outline
Professional Science-Based Work
Natural Resources Management and Biological Sciences (400 Group)
    The Forestry Profession
    The Range Manager Profession
    The Fisheries Manager Profession
    The Wildlife Manager Profession
Related Professions
    General Natural Resources Management and Biological Sciences (401 Series)
    Professional Work Compared to Technical Jobs
    Range Technician (455 Series)
    Forest Technician (462 Series)
    Park Ranger Profession (25 Series)
    Wildland Fire
Federal Tours of Duty and Position Types
Federal General Schedule Education Requirements and Salaries

The professions of forester, wildlife manager, fisheries manager, and range manager are the primary focus of this book. These are the traditional natural resource, or conservation, professions that have developed in the United States over the past 120 years. Although they are all natural resource professions, they have been taught as separate disciplines for decades. Each has developed distinct skill sets and spheres of influence; however, the basic processes used for the management of the resources are similar. The process starts with measuring the health of the resource. Remember the saying from Chapter 1: if you can't measure it, you can't manage it. Measuring the resource is the precursor to managing it. Measurement data is then analyzed and interpreted to inform decisions on how the resource should be managed. The basic process and fundamental science may be the same for each of these professions, but the advanced knowledge and skill sets differ. Members of these professions also must communicate and interface with different, but overlapping, constituencies or groups of people interested in the particular resource they are managing.

As mentioned in Chapter 1, the term *wildlife management* will be used to describe both wildlife biology and wildlife management and, to remain consistent, the term *fisheries management* will be used to describe both fisheries science and management. The term *traditional natural resource profession* will be used to describe all four professions. There will be a section about each that begins with a definition of the profession and includes a description of the job and its requirements. Additional natural resource professions will also be described because they are either closely related to, or often confused with, forestry, wildlife management, fisheries management, and range management. These related professions are natural resource scientist, range technician, forestry technician, park ranger, and wildland firefighter.

This chapter also includes a description of federal terms of service and pay rates for professional positions. The discussion of all the professions will be weighted heavily toward federal job descriptions and requirements. There are two reasons for this. First, the federal government is a dominant employer of natural resources professionals. Second, the federal government's job descriptions and requirements are consistent and readily available on agency websites, and these descriptions are broad enough to encompass most of the duties of jobs for other government and nongovernment entities within these professions.

The information in the following discussion was adopted from the federal *Office of Personnel Management List of Occupational Standards* (Office of Personnel Management 2017a), *Group Standards* (Office of Personnel Management 2017b), and their *Professional Work in the Natural Resources Management and Biological Sciences Group, 0400* (Office of Personnel Management 2005). Rather than clutter the text and cite these sources repeatedly throughout the chapter, the convention will be to cite only information that did not come from one of these three sources.

PROFESSIONAL SCIENCE-BASED WORK

The management of forests, wildlife, fisheries, and rangeland requires individuals who are professionals. These managers are members of a professional community with these characteristics: (1) members are required to have training at the college level, (2) each profession has a professional association, (3) each has a code of ethics, and (4) those practicing within the field are required to have some type of licensing or certification. The federal government's definition of **professional scientific work** is that which involves exercising discretion, using analytical skills and judgment, and requiring personal accountability and responsibility for creating, developing, integrating, and applying an organized body of knowledge. This knowledge is acquired through extensive education, or training, at a recognized college or university equivalent to the curriculum requirements for a bachelor's degree or higher, with major study in or pertinent to the specialized field. Moreover, the field is continuously studied to explore, extend, and use additional discoveries, interpretations, and applications to improve data quality, materials, equipment, applications, and methods.

What that means in common terms is that a natural resources professional will need a four-year college degree with college coursework in the specific topics of the profession. The professional will be given a large amount of responsibility and is expected to be able to collect data and information, analyze and interpret this data, and then exercise discretion and judgment to apply the interpretation to the administration and management of the resource. Early in their careers, professionals will likely be involved mainly in the collection of data and information. If they demonstrate good critical thinking skills and can analyze and interpret the data and information, they will progress to managing the resource. If they further demonstrate good social skills, judgment, discretion, and communication competence, they will progress to administration of others who manage the resource. Along this career path such professionals are expected to continue gaining knowledge and skills through formal continuing education programs.

This is a general definition applicable to all science-based professional work, but there are differences in objectives between government and private organizations. Foremost among these is that private companies make economic gain a high priority. So although a natural resources professional working for a private company will be applying knowledge and skills similar to that applied by government employees, the objectives and desired outcomes of their work will differ. In fact, objectives will vary not only between but also *within* agencies and organizations. However, the skills are largely the same, and some professionals move back and forth between private and government jobs during their careers.

Natural Resources Management and Biological Sciences (400 Group)

The US government's Office of Personnel Management (OPM) classifies federal jobs into groups and numbered series. The 400 job group is the Natural Resources Management and Biological Sciences Group. Within this group, the jobs deal primarily with water, land, food, plants, animals, and soils, and may involve managerial or administrative duties. The federal description of the work is that it requires the interpretation and application of biological science and research as well as the application of management concepts and practices. In addition to the biological sciences, the work may encompass aspects of the social, behavioral, cultural, and economic sciences as applied to conserving, utilizing, and sustaining natural, physical, and cultural resources of forests, land, and water.

Broad examples of the work of a natural resources professional employed by a federal agency are:

- protecting sensitive habitats
- advising on natural resources regulations
- ensuring that activities such as hunting and boating are conducted lawfully
- developing environmental reports and impact statements
- developing and implementing land-use plans
- identifying problems associated with public land use, such as mining and drilling operations, utility corridors, road and highway development, or telecommunications facilities
- coordinating the surveying, monitoring, analyzing, and evaluating of natural resources to permit multiple uses while preserving the area's ecological viability
- managing special areas, such as wilderness areas, recreation areas, and trails
- managing wildlife and fisheries
- protecting threatened and endangered species and sensitive plants
- managing wildland fire programs
- managing recreation programs

A forester samples foliage for spruce bud worms. (Photo by USFS, Flickr)

Natural resource jobs require a breadth of knowledge and a variety of skills. Depending on the employer's objectives, most of the examples given can be applied to professionals working for state or local governments, private companies, or other organizations.

The Forestry Profession

Forestry, as defined in Chapter 1, is the profession embracing the science, art, and practice of creating, managing, using, and conserving forest and associated resources for human benefit in a sustainable manner to meet desired goals, needs, and values. The broad field of forestry consists of those biological, quantitative, managerial, and social sciences that are applied to forest management and conservation.

Within the wider field of forestry, there are several types of foresters, although these definitions vary somewhat between states and regions of the country. *Forest managers* specialize in managing forests. *Service foresters* are agency foresters who advise private landowners with forestry issues. *Forestry extension agents* assist private landowners as part of land-grant university systems. *Compliance foresters* inspect, assist, and enforce forest regulations. *Utility foresters* work for or assist utility companies in managing vegetation in powerline and pipeline corridors. *Area foresters* are responsible for managing a large land area and are often supervisors of several other foresters. *Forest engineers*, some of whom are also licensed civil engineers, design forest harvest systems and forest road systems, including drainage and bridges. *Operations foresters* design and manage harvesting operations. *Recreation foresters* manage forest areas likely to be highly impacted by recreation. *Silvicultural foresters*, or silviculturists, are responsible for the growth and tending of forests. *Inventory foresters* are responsible for measuring, mapping, and accounting for the growth of a forest. *Wood scientists* are often foresters who specialize in wood-product manufacturing and quality control.

Professional Associations. The dominant professional society is the Society of American Foresters (SAF) founded in 1900. The SAF provides accreditation to colleges, certification to individual foresters, and continuing education opportunities for foresters. Details of accreditation and certification can be found in Chapter 5 and Appendix A and B. Other professional organizations of foresters are the Association of Consulting Foresters and the Forest Stewards Guild.

Responsibilities. The federal government's description of the general duties of a forester includes:

1. managing forestlands and grazing areas
2. producing timber
3. conserving soil
4. preserving wildlife and wildlife habitats
5. protecting watersheds
6. developing recreational sites and opportunities

Foresters develop, oversee, and protect forests. Forestry work involves inventorying, planning, evaluating, and managing forest resources that include timber, soil, land, water, wildlife and fish habitats, minerals, forage, and outdoor recreation. Foresters are also responsible for protecting resources against fire, insects, disease, floods, and erosion. Foresters may be involved in (1) interpreting and communicating legislation about forestland management; (2) applying principles of sustained yield management to forest resources, wetlands, water and soil quality, and wildlife conservation to protect forested lands during timber harvesting operations; and (3) developing new, improved, or more economic scientific methods, practices, or techniques necessary to perform such work.

Additionally, foresters cooperate with and provide technical assistance to others; conduct research on managing timber, forest watersheds, and other related resources; appraise forestlands, properties, and resources for acquisition, taxation, sale, or exchange; evaluate policy issues and environmental regulations that affect many forestry-related activities; and recommend policies and programs to keep forestlands productive. Foresters prepare both short- and long-term land management and multi-resources plans with alternative strategies, which involves analyzing the environmental, economic, and social effects of each alternative. Foresters prepare these plans in consultation with others, including the public.

Employment Statistics. The federal designation for a professional forester is the Forestry 460 series. There are about 8,300 foresters in the United States and in 2017 they earned a mean annual salary over $61,000 (Table 2-1). Most US foresters work for a government agency, with various state agencies employing the largest numbers. Interestingly, utility foresters working for power distribution companies have the highest mean annual incomes. There is projected to be about a 5 percent expansion in the number of foresters employed by 2026, with about 1,100 openings for foresters between 2016 and 2026 (Table 2-2).

Table 2-1 Employment of Foresters in the United States, 2017*

Foresters	Number of Jobs	Mean Annual Salary ($)
State governments	3,010	57,790
Federal executive branch	1,270	65,020
Local governments	1,100	61,520
Sawmills and wood preservation industry	580	62,500
Logging industry	580	64,470
Electric power generation, transmission and distribution	300	75,310
Management of companies and enterprises	200	70,630
Pulp, paper, and paperboard mills	160	68,980
Veneer, plywood, and engineered wood product manufacturing	100	68,930
Other	1,000	
Total	**8,300**	**61,710**

Note: Does not include self-employed individuals or owners/partners in unincorporated firms. (*Source:* Bureau of Labor Statistics 2018a.)

Table 2-2 Employment Projections for US Foresters

Occupation	2016 Jobs	2026 Jobs	Employment Change in Jobs	Employment Change (percent)	Potential Openings 2016 to 2026*	2017 Median Wages ($)
Forester	12,300	12,900	600	5.0	1,100	60,120

Note: Potential openings are greater than employment change due to retirements and people leaving the profession.
(*Source*: Bureau of Labor Statistics n.d.)

Academic Requirements. The basic requirement for a federal forester's position is a bachelor's degree in forestry, or a related field, that includes at least 30 semester hours of biological, physical, or mathematical sciences or engineering, of which at least 24 semester hours are coursework in forestry. The curriculum must include courses in each of the following areas:

1. *Management of Renewable Resources*—the study of the science and art of managing renewable resources to attain desired results. Examples of courses in this area include silviculture, forest management operations, timber management, wildland fire science or fire management, utilization of forest resources, forest regulation, recreational land management, watershed management, and wildlife or range habitat management.
2. *Forest Biology*—the study of the classification, distribution, characteristics, and identification of forest vegetation, and the interrelationships of living organisms to the forest environment. Examples of courses in this area include dendrology, forest ecology, silvics, forest genetics, wood structure and properties, forest soils, forest entomology, and forest pathology.
3. *Forest Resource Measurements and Inventory*—the study of sampling, inventory, measurement, and analysis techniques as applied to a variety of forest resources. Courses include forest biometrics, forest mensuration, forest valuation, statistical analysis of forest resource data, renewable natural resources inventory and analysis, and photogrammetry or remote sensing.

The federal requirements allow applicants to qualify as a forester through a combination of education and experience when courses are equivalent to a major in forestry and they have appropriate experience. An applicant may qualify as a Forester (Administration) or Research Forester (Administration) through the educational requirements established for other forestry-related professional disciplines as long as they have a sufficient amount of professional experience gained in a forestry work situation.

State Licensing. Currently, thirteen states require licensing or certification of foresters working in their state (Table 2-3). The term Registered Professional Forester (RPF) is used by many states to denote a licensed forester. Licensing requirements can be quite extensive. For example, California requires seven years of experience working under the supervision of a licensed forester, some of which may be offset with approved college education; reference checks for character; and passage of a comprehensive exam before granting an individual the status of California Registered Professional Forester.

Table 2-3   State Boards that Oversee Licensing of Foresters

Alabama State Board of Registration of Foresters
Arkansas Board of Registration for Foresters
California Board of Forestry and Fire Protection
Connecticut Department of Energy and Environmental Protection
Georgia State Board of Registration for Foresters
Maine Board of Licensure of Foresters
Maryland State Board of Foresters
Massachusetts Forester Licensing Board
Mississippi State Board of Registration for Foresters
New Hampshire Board of Foresters
North Carolina State Board of Registration for Foresters
South Carolina State Board of Registration for Foresters
West Virginia State Board of Registration for Foresters

(*Source*: Society of American Foresters 2017.)

The Range Manager Profession

Range management "is a distinct discipline founded on ecological principles that deals with the use of rangelands and range resources for a variety of purposes" (Idaho OnePlan n.d.). These purposes include the use of rangelands as watersheds, wildlife habitat, and livestock grazing areas; and for recreational and aesthetic uses. Range management specialties include assessing and monitoring rangelands, rangeland conservation, hydrology, rangeland management, and rangeland administration.

Professional Associations. The Society for Range Management (SRM), founded in 1948, is the professional organization that represents range managers in the United States. SRM provides accreditation of college range management programs, certification of individual range managers and consultants, and continuing education in range management.

Responsibilities. The federal government's description of the duties of a rangeland manager includes:

1. analyzing and protecting natural resources
2. developing programs and standards for rangeland use and preservation
3. advising officials and landowners on rangeland management practices

Professions

The Bureau of Land Management (BLM) manages public rangelands for various uses, including livestock grazing. (Photo by BLM, Flickr)

Rangelands include grasslands, savannas, shrublands, riparian properties, pastures, hay lands, deserts, tundra, alpine communities, coastal marshes, and wet meadows. Rangeland management has a large ecological component and range managers must apply a knowledge of science, such as plant, animal, and soil sciences; watershed, habitat, and wildlife management; ecology; animal husbandry; economics; hydrology; agronomy; soil conservation and management; livestock management; recreation management; and forestry.

Range managers provide technical recommendations on managing public and private rangelands for ecological improvement consistent with objectives set forth in land-use planning documents. They manage rangelands and their various resources to meet the present and future needs of the public. These resources include vegetation, soil, water, timber, minerals, wildlife habitats, historic and prehistoric resources, wilderness, scenery, open space, and a rural way of life. Use of rangelands include:

- livestock grazing
- wildlife habitat
- recreation
- water
- timber production

- mineral development
- producing forage for domestic and wild animals
- protecting threatened and endangered plant and wildlife species

Range managers prepare both short- and long-term land-use plans that include analysis of the environmental, economic, and social effects of the proposed alternatives. The plans are prepared with public input from diverse interest groups that often have diametrically opposed goals and objectives. Other responsibilities of a range manager include: developing conservation plans, designing technical surveys, and supervising construction; developing contractual agreements between agencies and private landowners and/or contractors; protecting cultural resources; and working with other federal, state, and local conservation agencies.

Employment Statistics. The federal designation for a range manager is the Rangeland Management 454 series. Unfortunately, there are no employment statistics for range managers because government employment agencies do not have a distinct category for range management.

Academic Requirements. The basic requirement is a degree in range management or a related discipline that includes at least 42 semester hours in a combination of plant, animal, and soil sciences; and natural resources management, as follows:

1. *Range Management*—at least 18 semester hours of coursework in range management, including courses in the basic principles of range management, range plants, range ecology, range inventories and studies, range improvements, and ranch or rangeland planning.
2. *Directly Related Plant, Animal, and Soil Sciences*—at least 15 semester hours of directly related courses in the plant, animal, and soil sciences, including at least one course in each of these three scientific areas. Courses in such areas as plant taxonomy, plant physiology, plant ecology, animal nutrition, livestock production, and soil morphology or soil classification are acceptable.
3. *Related Resource Management Studies*—at least 9 semester hours of coursework in related resource management subjects, including courses in such areas as wildlife management, watershed management, natural resource or agricultural economics, forestry, agronomy, forages, and outdoor recreation management.

The requirements allow an applicant to qualify as a range manager through a combination of education and experience, with at least 42 semester hours of coursework in natural resources management and plant, animal, and soil sciences, plus appropriate experience.

The Fisheries Manager Profession

The American Fisheries Society (AFS) defines fisheries as "the career field dealing with fisheries science and management." A *fishery* is an aquatic resource that has biological, environmental, and social components (American Fisheries Society n.d.). Fisheries professionals may work for the government, nonprofit organizations, private industry, or universities or colleges, and their work may range from conducting research and investigating problems, to managing a specific fishery in a region, to raising aquatic species in a hatchery. Some of the jobs that are available in fisheries include: fisheries manager, fisheries researcher, hatchery manager, or consultant.

Professional Associations. The American Fisheries Society, founded in 1870, is the dominant professional association and provides individual certification and continuing education opportunities for fisheries managers. Certification will be explained in detail in Chapter 5. Other professional societies applicable to fisheries are the American Society of Limnology and Oceanography and the American Institute of Fisheries Research Biologists.

Responsibilities. There are several different job categories for fisheries managers within the federal 400 job group. The federal designation for a fish or wildlife administrator is the Fish and Wildlife Administration 480 series. Fish and wildlife administration work involves

Fish biologists collect data for fish habitat studies on the Deschutes National Forest. (Photo by USFS, Deschutes National Forest, Flickr)

conserving, enhancing, and protecting fish, wildlife, plants, and their habitats. Responsibilities for the positions include advocacy and leadership in administering and managing fish and wildlife resources as required by legislation. The work of a fish or wildlife administrator involves directing and formulating policies, plans, standards, and procedures for comprehensive fish and wildlife conservation and restoration programs, as well as coordinating grants-in-aid for fish and wildlife programs.

Fish and wildlife administrators establish, maintain, and nurture relationships with state agencies, conservation organizations, other federal agencies, and the fishing, hunting, and boating industries. They are involved in negotiating with state fish and wildlife officials on activities allowable under federal natural resources programs and provide expertise to assure compliance with laws, regulations, policies, and executive orders applicable to natural resources programs and activities. Their duties also include interpreting and implementing legislation and legal decisions impacting natural resources practices and programs; advising agency officials on complex scientific, political, and economic natural resources issues; and auditing federal grant programs to determine continuing fund eligibility.

The federal designation for what is described in this book as a fisheries manager is the 482 Fish Biology series. Fish biology work ranges from directly managing fish resources to studying and analyzing fish life history, behavior, habitat requirements, classification, and economic implications. Managing fish resources involves:

1. assessing and mitigating environmental and human impacts on the survival and growth of aquatic species and their habitats
2. operating physical facilities and equipment
3. analyzing and planning for secure sustained yields and species' long-term survival and contribution to ecosystem functions
4. coordinating with other natural resources activities, such as forest management, range management, and land-use planning

The fisheries manager considers the conservation, culture, nutrition, fish health, and habitat restoration of fish and other aquatic species (crabs, shrimp, or oysters) in the context of their roles in the ecosystem. Research work may involve studying various ecological systems in relation to the health, growth, and well-being of fish resources. It may also include conducting surveys, designing and implementing restoration plans, and developing recovery plans and other fish management plans; identifying and protecting aquatic habitats and associated and interconnected uplands that contribute to stream and lake habitat quality; developing methods to culture fish (for example, hatchery operations), and dealing with fish health issues.

Professions

Employment Statistics. Since employment statistics for fisheries managers, wildlife managers, and zoologists are combined into a single category by the Bureau of Labor Statistics, we cannot provide employment numbers for fisheries managers alone. As of May 2017, there were nearly 18,000 wildlife managers, fisheries managers, and zoologists working in the United States, with a mean annual salary of more than $66,000 (Table 2-4). The dominant employer is various state governments; the federal government is the second largest employer. The projected growth in employment between 2016 and 2026 in these professions is 7.6 percent, with an estimated 1,900 positions opening during this time span (Table 2-5).

Table 2-4 Employment of Wildlife Managers, Fisheries Managers, and Zoologists in the United States, 2017*

Wildlife, Fisheries, and Zoology	Number of Jobs	Mean Annual Salary ($)
State government	6,650	57,190
Federal executive branch	4,120	82,490
Management, scientific, and technical consulting services	1,410	75,840
Local government	1,320	64,370
Colleges, universities, and professional schools	1,270	61,860
Scientific research and development services	900	63,650
Social advocacy organizations	550	62,570
Architectural, engineering, and related services	430	64,090
Management of companies and enterprises	110	80,240
Other	950	
Total	**17,710**	**66,250**

* *Note:* Does not include self-employed individuals.
(*Source:* Bureau of Labor Statistics 2018b.)

Table 2-5 Employment Projections for Wildlife Managers, Fisheries Managers, and Zoologists

Occupation	2016 Jobs	2026 Jobs	Employment Change in Jobs	Employment Change (percent)	Potential Openings 2016 to 2026*	2017 Median Wages ($)
Wildlife, Fisheries, and Zoology	19,400	20,900	1,500	7.6	1,900	62,290

* *Note:* Potential openings are greater than employment change due to retirements and people leaving the profession.
(*Source*: Bureau of Labor Statistics n.d.)

Academic Requirements. The basic federal requirement for a fish or wildlife administrator is a bachelor's degree appropriate to the position. An alternative to this requirement is a combination of education and experience with courses equivalent to a major, or at least 30 semester hours in biological sciences, agriculture, natural resource management, chemistry, or related disciplines, plus appropriate experience.

The basic requirement for a federal fisheries manager in a non-research position is a degree with a major in biological science that includes at least 6 semester hours in aquatic subjects such as limnology, ichthyology, fishery biology, aquatic botany, aquatic fauna, oceanography, fish culture, or related courses in the field of fish biology; and at least 12 semester hours in the animal sciences in such subjects as general zoology, vertebrate zoology, comparative anatomy, physiology, entomology, parasitology, ecology, cellular biology, genetics, or research in these fields. A combination of education and experience that includes courses equivalent to a major in biological science (i.e., at least 30 semester hours), of which a minimum of 6 semester hours are in aquatic subjects and 12 semester hours are in the animal sciences, plus appropriate experience, will also meet the federal requirement for this position.

The basic requirement for a federal research position in fish biology is a degree with major studies in biology, zoology, or biological oceanography that includes at least 30 semester hours in biological and aquatic science and 15 semester hours in the physical and mathematical sciences. This coursework must include at least 15 semester hours of preparatory training in zoology beyond introductory biology or zoology in such courses as invertebrate zoology, comparative anatomy, histology, physiology, embryology, advanced vertebrate zoology, genetics, entomology, and parasitology; and at least 6 semester hours of training applicable to fishery biology in such subjects as fishery biology, ichthyology, limnology, oceanography, algology, planktonology, marine or fresh water ecology, invertebrate ecology, principles of fishery population dynamics, or related course work in the field of fishery biology; and at least 15 semester hours of training in any combination of two or more of the following: chemistry, physics, mathematics, or statistics.

The Wildlife Manager Profession

Wildlife management is defined as "the management of wildlife populations within the context of the ecosystem" (Fryxell et al. 2014). These authors state that wildlife management is the manipulation of a population to achieve a goal and that wildlife management implies stewardship or, if stewardship fails, conservation. They also explain that management may be custodial, a preventative or protective role, or manipulative using either direct control of the populations or indirect control by changing the population's environment.

Definitions and Responsibilities. The Wildlife Society (n.d.) gives the following definitions of wildlife occupations.

1. *Wildlife Manager*—maintains or manipulates wildlife populations, habitats, or human users to produce benefits for wildlife and the general public. Benefits sought may be ecological, economic, social, recreational, or scientific. A wildlife manager uses wildlife science to formulate and apply scientifically sound solutions to wildlife and habitat management problems.
2. *Wildlife Biologist*—gathers, analyzes, and interprets data on wildlife and habitats, including behavior, disease, ecology, genetics, nutrition, population dynamics, physiology, land-use changes, and pollution to conserve wildlife species and improve habitat conditions. A wildlife manager uses scientific principles to research wildlife and habitats to increase our knowledge base.
3. *Wildlife Educator*—teaches high school and university students about wildlife science and conservation including wildlife biology, ecology, physiology, disease, toxicology, taxonomy, economics, research and management techniques, and conservation policy and law.
4. *Wildlife Public Educator and Outreach Specialist*—educates the public about wildlife species and conservation issues. Outreach specialists work with private and corporate landowners, industries, citizen groups, and others to provide technical assistance related to wildlife management on private or public farms, forests, parks, urban areas, and industrial lands. Outreach specialists apply economic principles and conservation practices to aid others in maintaining or restoring wildlife on their lands.
5. *Wildlife Law Enforcement Officer*—enforces wildlife laws and regulations to maintain wildlife populations at desired levels. Wildlife law enforcement officers often perform surveys of wildlife populations, are involved in trapping and banding programs, implement wildlife population controls, respond to complaints of nuisance wildlife, and educate the public about wildlife issues.
6. *Wildlife Technician*—collects data on wildlife and habitats under the supervision of a Wildlife Manager.
7. *Wildlife Inspector and Forensics Specialist*—intercepts smuggled, illegal shipments of live wild animals for the pet trade and wild animal parts for trophy or medicinal purposes. Wildlife inspectors are stationed at international airports, ocean ports, and border crossings. Forensics specialists perform scientific and investigative work to document the origin and nature of evidence collected on these illegal imports.

8. *Wildlife Communications and Public Relations Specialist*—interprets wildlife research and conservation programs to present to the general public. Communications and public relations specialists write articles and news releases, create brochures and websites, photograph wildlife and conservation activities, and speak at public gatherings or through the media.
9. *Wildlife Policy Analyst*—applies wildlife management theories and practices to laws and regulations governing wildlife and habitats. Wildlife policy analysts often work for governments, legislative bodies, nonprofit organizations, or industry groups.
10. *Wildlife Consultant*—evaluates ecosystems to determine environmental impacts from proposed actions. Following standards created by the National Environmental Policy Act, consultants provide reports to businesses, industries, and governments to ensure quality environments.
11. *Wildlife Economist*—provides economic analyses of natural resources to support policies, critical habitat designation, assess damage, and analyze environmental plans.
12. *Wildlife Administrator*—works with many stakeholders and budgets to provide assistance in promoting sound resource management programs designed to effectively manage wildlife and habitats.
13. *Wildlife GIS Specialist*—works with geographic information systems and other technologies to interpret data and make management and policy recommendations concerning wildlife and their habitats.

The federal designation for what this book describes as a wildlife manager is the Wildlife Biology 486 series. The federal definition of a wildlife manager is a professional who deals with the ecology, behavior, and conservation of wild animals and coordinates wildlife management programs with other natural resources activities, such as land-use planning and forest and range management. The management work includes:

1. developing and managing wildlife programs on federally owned or managed lands, such as national parks, national forests, wildlife refuges, Indian reservations, military installations, wetlands, big game and desert ranges, and other lands in the public domain
2. developing and implementing cooperative programs with and providing technical assistance to states, private landowners, Alaskan Native and Indian tribal governing bodies, and special interest groups concerned with protection and proper management of wildlife and wildlife habitats

A US Fish and Wildlife Service wildlife biologist radio tracking a pair of mountain lion siblings. (Photo by Carman Luna, USFWS, Flickr)

3. assessing and conducting wildlife management transactions, such as acquiring, selling, leasing, or exchanging lands, easements, and other resources
4. preparing, evaluating, and conducting biological analyses of land and water resources projects and federal permit applications to ensure compliance with appropriate law and to mitigate adverse impacts on resources
5. reviewing state and federal proposals for funding wildlife resources projects to determine if planned objectives warrant federal funding and meet wildlife resources needs in accordance with applicable laws and regulations

The research work of a wildlife manager includes proposing, designing, and conducting studies to determine: population status, trends, and problems of wildlife species; disease control specifications; endangered or threatened species protection and consultation requirements; planned habitat management actions and evaluation procedures; population enhancement programs; and environmental contaminant specifications.

The federal designation for a wildlife manager who manages a wildlife refuge is the Wildlife Refuge Management 485 series. The wildlife refuge manager is a person involved with developing, enhancing, protecting, and maintaining land and habitat for a variety of species within the confines of a wildlife refuge. The difficulty of the job differs among refuges—depending on the species involved, the required protection, public use, commercial interests, water supply, and the interests of other government agencies. Wildlife refuges vary in size, topography, geographic location, climate and other characteristics. Physical characteristics may include arctic tundra, desert, bog and marshlands, estuarine, coastline, wetlands, and uplands. Refuges may be pristine, or contain inhabited communities and historical landmarks.

The work of a refuge manager involves:

- planning land, water, and habitat management
- administering, supervising, and managing fish and wildlife public relations activities
- managing public, commercial, industrial, and agrarian land use
- preserving, restoring, and enhancing populations of endangered or threatened species of animals and plants
- perpetuating migratory and residential bird resources
- preserving a natural diversity and abundance of fauna and flora

The refuge manager is also responsible for providing the public with an understanding and appreciation of fish and wildlife ecology; providing the public with recreational opportunities such as nature trails, hunting, fishing, and observation; studying the characteristics and behavior of species; evaluating the adequacy of habitats to support wildlife needs; evaluating administration practices for one species relative to its impact on other species and their habitats; identifying and applying disease control and containment methods; ensuring that public uses are authorized and compatible with the purposes for which a refuge was established; and preventing adverse impacts on wildlife species and national historic sites.

The duties of a refuge manager will also involve contracting for business operations and issuing permits for economic uses of resources, such as farming, mineral exploration and extraction, and power production; reconciling biological program compatibility with other needs and activities in surrounding communities; and assessing the impact of agricultural and commercial activities or military operations on nearby managed property. Administrative aspects of the work may require an understanding of the basic principles, concepts, and techniques of budgeting, contracting and procurement, personnel, records management, and property management.

Professional Associations. The Wildlife Society, founded in 1937, offers individual certification and continuing education for wildlife managers. Read more about certification in Chapter 5. Other professional societies relevant to wildlife are the American Ornithological Society, the Association of Field Ornithologists, the American Society of Ichthyologists and Herpetologists, the American Society of Mammalogists, the Herpetologists League, the Society for Marine Mammalogy, and the Society for the Study of Amphibians and Reptiles.

Employment Statistics. Since employment statistics for wildlife managers, fisheries managers, and zoologists are combined into a single category by the Bureau of Labor Statistics, there is no employment data for wildlife managers alone. As of May 2017, as a group, there were nearly 18,000 wildlife managers, fisheries managers, and zoologists working in the United States, with a mean annual salary of more than $66,000 (Table 2-4). The dominant employer is various state governments, with the federal government being the second largest employer. The projected growth in employment between 2016 and 2026 in these professions is 7.6 percent, with an estimated 1,900 positions opening during this time span (Table 2-5).

Academic Requirements. The basic requirement for a nonresearch wildlife manager is a degree in biological science that includes: at least 9 semester hours in such wildlife subjects as mammalogy, ornithology, animal ecology, wildlife management, or research courses in the field of wildlife biology; at least 12 semester hours in zoology in such subjects as general zoology, invertebrate zoology, vertebrate zoology, comparative anatomy, physiology, genetics, ecology, cellular biology, parasitology, entomology, or research courses in such subjects; and at least 9 semester hours in botany or the related plant sciences, or a combination of education and experience.

The basic requirement for a research wildlife manager is a degree in wildlife biology, zoology, or botany that includes at least 30 semester hours of coursework in biological science and 15 semester hours in the physical, mathematical, and earth sciences. This coursework must include: at least 9 semester hours of training applicable to wildlife biology in such subjects as mammalogy, ornithology, animal ecology, wildlife management, principles of population dynamics, or related course work in the field of wildlife biology; at least 12 semester hours in zoological subjects such as invertebrate zoology, vertebrate zoology, comparative anatomy of the vertebrates, embryology, animal physiology, entomology, herpetology, parasitology, and genetics; at least 9 semester hours in the field of botany and related plant science; and at least 15 semester hours of training in any combination of two or more of the following: chemistry, physics, mathematics, statistics, soils, and/or geology.

The basic requirement for a federal job as wildlife refuge manager is a degree in zoology, wildlife management, or an appropriate field of biology that includes at least 9 semester hours in zoology; 6 semester hours in such wildlife courses as mammalogy, ornithology, animal ecology, or wildlife management; 3 semester hours in botany; and 3 semester hours in conservation biology, or a combination of education and experience.

Related Professions

Although the focus of this book is on the four traditional natural resources professions, this section provides a brief discussion of related occupations for comparative purposes.

General Natural Resources Management and Biological Sciences (401 Series)

It is important to contrast the professions of forester, range manager, wildlife manager, and fisheries manager with the broader profession that the federal government labels as General Natural Resources Management and Biological Sciences, or what will be called *natural resource scientist* within this book. Natural resource science is a broad discipline that developed later than forestry, wildlife management, fisheries management, and range management. (A brief history of the conservation movement and the development of these professions are in the next chapter of this book.) The profession of natural resource scientist is a newer profession and encompasses a wide range of possible jobs.

The federal description for the General Natural Resources Management and Biological Sciences (natural resource scientist) job is one that manages, supervises, leads, or performs professional research, or scientific work in biology, agriculture, or natural resources management *that is not classifiable to another more specific professional series* in the Natural Resources Management and Biological Sciences Group (400). Jobs classified within this general (401) series involve professional work in more than one series, or in the 400 group not covered by a specific series.

The basic requirement for a natural resource scientist is a bachelor's degree in biological sciences, agriculture, natural resource management, chemistry, or related disciplines appropriate to the position. A combination of education and appropriate experience with courses equivalent to a major in the above disciplines is also acceptable.

The 401 series is a catch-all description that is used when a position spans more than two of the other jobs described in the series. Most of the professional occupations in the 400 job group have requirements that also meet the criteria for the 401 series.

Professional Work Compared to Technical Jobs

The job descriptions and requirements outlined earlier with regard to the four traditional occupations primarily relate to professional positions. Technical positions, on the other hand, have different descriptions and requirements even though some tasks are common to both. The developmental work of professional positions and the work of high-level technical positions are sometimes similar and overlapping. In other words, the first few years of professional fieldwork in natural resources will often be identical to that of a technician, particularly that of an advanced technician. For example, the State of Oregon Employment Department (n.d.) indicates a 37 percent overlap in the skills of a forester with those of a forest technician. The federal distinctions between technical and professional work are given below.

Technical work involves the use of recurring methods, standardized procedures, and established processes for a specialized field, and applying knowledge acquired through practical experience and on-the-job activities with recognized processes, standards, and methods. Technicians must understand and apply predetermined procedures, methods, and standardized practices or approaches in a specialized field of industry or science. Technicians carry out tasks, procedures, and computations based on oral instructions and precedents, guidelines, and standards; and collect, observe, test, and record factual and scientific data with the oversight and management of professional employees. The work encompasses the ability to foresee the effects of procedural changes or to appraise the validity of results based on experience and practical reasoning, and the willingness to learn new practical methods and applications through on-the-job and classroom training. In summary, technical work (1) offers the opportunity to apply practical knowledge and skills, (2) requires the knowledge of practices, methods, and standards by which natural resources are made useful, and (3) supports and is associated with an industry, technology, or a professional scientific field.

Professional work involves creating, exploring, evaluating, and sharing solutions for scientific problems, conditions, and issues. Professionals apply a range and depth of knowledge acquired specifically through an intensive learning regimen of the theories and assumptions of scientific knowledge and the principles and practices of a professional discipline. Professional-level work involves identifying, analyzing, advising, consulting, and reporting on scientific, theoretical, and factual data, conditions, and problems. Such analysis includes assessing and predicting the relationships and interactions of data and findings under varying conditions. It also involves reasoning from existing knowledge and assumptions in a professional field to unexplored areas and phenomena. Professional-level work requires staying abreast of scientific subjects, analyses, and proposals in the professional literature. In summary, professional work offers the opportunity to (1) apply professional knowledge and skills and (2) to explore, create, and extend solutions and applications of a discipline.

In other words, technical work utilizes established procedures, repetitive methods, and is done with greater supervision than professional work, while professional work requires greater knowledge and creativity. The level of education required for a technical position is about half that required for a professional position.

Range Technician (455 Series)

The work of a range technician relates to the conservation, regulation, and use of public or federally controlled lands for grazing, range research activities, and livestock ranch operations. Range technicians also participate in wildland fire control, prevention, or suppression work.

Education and Training. The following discussions refer to the General Schedule (GS), a job classification and pay schedule that covers the majority of the federal government's professional, technical, administrative, and clerical positions. Additional information on the General Schedule appears later in this chapter.

A GS-3 position requires the successful completion of one year of study beyond high school. This one year of study must include at least 6 semester hours in a combination of courses such as range management or conservation, agriculture, forestry, wildlife management, engineering, biology, mathematics, or other natural or physical sciences.

A GS-4 position requires the successful completion of two years of study beyond high school. This additional study must include at least 12 semester hours in any combination of courses such as forestry, agriculture, crop or plant science, range management or conservation, wildlife management, watershed management, soil science, natural resources (except marine fisheries and oceanography), outdoor recreation management, civil or forest engineering, or wildland fire science. No more than 3 semester hours in mathematics is creditable.

A GS-5 position requires the successful completion of a full four-year course of study leading to a bachelor's degree with major study in forestry, range management, agriculture, or a subject-matter field directly related to the position; or a course of study with at least 24 semester hours in any combination of courses such as those shown above for the GS-4 level position, with no more than 6 semester hours in mathematics creditable.

Professions 33

Forest Technician (462 Series)

This series includes all positions that primarily require a practical knowledge of the methods and techniques of forestry and other biologically based resource management fields. Forestry technicians provide practical technical support in forestry research efforts; in the marketing of forest resources; or in the scientific management, protection, and development of forest resources. The primary focus of the forestry technician's work is to support an organization's resource-management efforts, from intensive multiple-use natural resource management to specialties such as marketing timber. Regardless of the mission of an organization, forestry technicians are most commonly found in field units and are principally concerned with performing work supporting the implementation of projects and program goals. They may be specialists in specific forestry subject areas or in the techniques and practices associated with production-oriented work.

One author has referred to forestry technicians as the noncom (non-commissioned officer) of the woods, with responsibilities intermediate between those of the skilled worker and the professional forester (Wray 1971).

Forestry Technician is the authorized title for federal positions at grades GS-4 and above. There are also *Lead Forestry Technicians* and *Supervisory Forestry Technicians* for those who meet higher qualifica-

A technician marks a "leave" tree with orange paint. Forest technicians often have the task of selecting trees to be cut and those that are left standing, or leave trees. (USFS, Flickr)

Table 2-6 Employment of Forest and Conservation Technicians in the United States, 2017*

Forest and Conservation Technicians	Number of Jobs	Mean Annual Salary ($)
Federal executive branch	21,510	39,190
State government	4,880	38,700
Local government	2,230	39,410
Management, scientific, and technical consulting services	720	35,810
Social advocacy organizations	370	40,690
Colleges, universities, and professional schools	330	42,950
Electric power generation, transmission, and distribution	40	57,960
Other	490	
Total	**30,570**	**39,180**

* *Note:* Does not include self-employed individuals.
(*Source:* Bureau of Labor Statistics 2018c.)

tions. Some federal agencies may supplement the basic title by adding a parenthetical suffix, where necessary, to identify duties and responsibilities that reflect specific knowledge and skills required in the work. Examples are Forestry Technician (Timber/Silviculture) or Forestry Technician (Fire Suppression). There is also a category called Forestry Aid at the GS-3 level and below.

Employment Statistics. The Bureau of Labor Statistics lumps forest technicians and conservation technicians together in its employment statistics. There were more than 30,000 forest and conservation technicians employed in the United States in May 2017; about 70 percent of these worked for the federal government (Table 2-6). State and local governments employed about 23 percent of forest and conservation technicians in the United States. The trend in employment indicates a 3.8 percent increase in positions between 2016 and 2026, which translates into an estimated 4,000 openings for forest and conservation technicians in this same period (Table 2-7).

Park Ranger Profession (25 Series)

This series covers positions in which the duties are to supervise, manage, and perform work in the conservation and use of federal park resources. This involves functions such as (1) park conservation, (2) natural, historical, and cultural resource management, and (3) the development and operation of interpretive and recreational programs for the benefit of the visiting public. Duties characteristically include assignments such as:

Professions

Table 2-7 Employment Projections for Forest and Conservation Technicians

Occupation	2016 Jobs	2026 Jobs	Employment Change in Jobs	Employment Change (percent)	Potential Openings 2016 to 2026*	2017 Median Wages ($)
Forest and Conservation Technician	33,200	34,500	1,300	3.8	4,000	36,130

Note: Potential openings are greater than employment change due to retirements and people leaving the profession.
(*Source*: Bureau of Labor Statistics n.d.)

- forest and structural fire control
- protecting property from natural or visitor-related depredation
- disseminating general, historical, or scientific information to visitors
- demonstrating folk-art and crafts
- controlling traffic and visitor use of facilities
- enforcing laws and regulations
- investigating violations, complaints, trespass/encroachment, and accidents
- search and rescue missions
- management activities related to resources such as wildlife, lakeshores, seashores, forests, historic buildings, battlefields, archeological properties, and recreation areas

Park rangers may perform work in one function, such as interpretive services, visitor protection, or visitor services, or the duties may involve various combinations of these functions. The work of park rangers varies considerably from park to park and the seasonality of operations. The term *park* may mean a national monument; seashore; parkway; historical, military, natural, or urban park; lake; or other related area administered by the Department of the Army and the Department of the Interior. The term *resource* includes natural, historical, cultural, archeological, or other similar types of resources. Park programs or functions range from preserving wilderness to operating urban parks; from protecting natural forests and historical buildings to safeguarding people on crowded recreational beaches or lakes; from patrolling back country areas to delivering interpretive talks in parks, community centers, schools, and similar establishments; from fighting forest fires to controlling large crowds; from overcoming encroachments on public lands to encouraging people to properly use and enjoy park facilities. Generally, the work of park rangers falls into three broad functional areas:

1. *Interpretation*—which involves interpreting the natural, historical, archeological, or other features of the resource and area to enrich the visitors' experience through activities such as talks, guided or self-guided walks, campfire presentations, demonstrations, and environmental education programs both in the park and in community centers, schools, or other related nonpark locations.
2. *Visitor Protection and Service*—which involves activities such as operating campgrounds, marinas, picnic areas, and other recreation facilities; search and rescue or other emergency services; boat, road, or other patrol activities for enforcement and inspection purposes; traffic control; and fee collection.
3. *Management*—which involves protecting, managing, and conserving the natural, historical, and other characteristics of the area through activities such as forest, wildland, and structural fire prevention and suppression; boundary encroachment and land-use activities; fish and wildlife management; preservation of natural, cultural, and/or historical structures and objects; and flood control activities.

Park rangers at the GS-2 or GS-3 level typically acquire the basic knowledge, skills, and abilities needed to carry out their duties through on-the-job training and experience. These are often temporary or seasonal positions. Entry at higher grade levels may be gained through specifically related education and specialized experience.

Wildland Fire

Wildland firefighting consumes a significant portion of many natural resource agency budgets. Many foresters, wildlife managers, and fisheries managers are engaged in firefighting or have their management activities affected by a wildland fire or the efforts to limit wildland fires. Regardless of the type of work and whether the employee is permanent or seasonal, generally everyone who does fieldwork and can pass the medical and physical fitness requirements receives some basic wildland firefighter training. This is true within public agencies and private organizations. Many natural resources professionals are assigned to wildfire suppression or management when needed, even if it means keeping records or transporting people or equipment. Definitions are given below for some of the common wildland fire jobs and crews.

1. *Seasonal Firefighters.* To become a wildland firefighter, you must be between 18 and 35 years old and pass a physical fitness test. Most of the federal seasonal firefighter jobs are at the Forestry

Professions 37

A prescribed burn being used as live-fire training. Natural resources professionals are often used to supplement regular wildland fire suppression crews. Most field professionals will receive fire training from their company or agency. (Photo by author)

Aid (GS-3 and below) and Forestry Technician (GS-4 and above) level. States generally hire firefighters at similar pay rates. Each federal agency and state agency has its own process for hiring seasonal employees. Most agencies hire many employees on a seasonal basis primarily from May to September.

2. *Professional Full-Time Firefighters.* These workers manage fire programs—which involves advising, administering, supervising, or performing professional fire management work. The Office of Personnel Management (2005) describes this work as requiring knowledge of fire behavior, firefighting techniques, and wildland fire management. Typical fire program management activities are assessing potential and actual fire effects on riparian areas, soil, air quality, wildlife habitats, and cultural, commercial, and recreational resources; protecting and restoring ecosystems from fires; developing incident management strategies, tactics, and plans; implementing prescribed (controlled burn) fires; and planning and coordinating fire protection programs and fire responses with other federal, state, and local agencies as well as private interests. Many of these positions are forest technician level positions for either state or federal agencies.

Wildland Fire Crew Types. The following descriptions are adapted from those provided on the US Department of Interior Wildland Fire Jobs website (2017). Most of these crews will be composed of workers hired specifically for wildland fire positions and not those assigned temporarily to firefighting.

Engine Crews are used for initial and extended attack fire suppression, support of prescribed fires, patrolling, and project work. These crews range in size from 3 to 10 firefighters and work with specialized firefighting equipment. They perform many strenuous activities such as mobile attack with engines, hose lay, construction of fire line with hand tools, burnout operations, and mopping up hotspots.

Hand Crews normally consist of 18 to 20 members. These crews can be used for a variety of operations. Hand crews are assigned duties on wildland and prescribed fires that primarily consist of constructing fire lines with hand tools and chainsaws, burning out areas using drip torches and other firing devices, and mop-up and rehabilitation of burned areas. Hand crews may or may not have assigned permanent supervision.

Helitack Crews are specialized wildland fire-suppression crews trained in helicopter operations. Helitack firefighters are delivered to fires via helicopter and suppress wildfires with hand tools and chainsaws. Some helitack firefighters are trained to rappel from the helicopter to reach fires in remote locations. The crew can range in size from 7 to 10 people. Helitack crews may also be used to support prescribed fire operations or special projects requiring helicopters.

Interagency Hotshot Crews are 20-person, highly skilled, experienced crews used primarily for wildfire suppression and fuels reduction. They perform the same duties as other hand crews, but are very mobile and generally placed in the most rugged terrain on the most active and difficult areas of a wildfire. Hotshot crews are considered "national assets" and spend extended periods away from their home units and travel throughout the country. The crews place a great deal of emphasis on physical fitness. There were over 110 hotshot crews nationwide in the summer of 2017.

Smokejumpers are highly trained and experienced firefighters who parachute into areas of wildfire to provide a rapid response to an emerging or ongoing fire. A plane load of smokejumpers can vary from 8 to 20 people, depending on the aircraft. All smokejumpers must have previous firefighting experience and have the most stringent physical fitness requirements of any wildland firefighters.

Fuels Crews are used primarily for working on fuels projects, which include hazardous fuels reduction and restoration of fire adapted ecosystems. This work may entail thinning of timber or shrubs with chainsaws, conducting controlled burns, pilling and chipping of slash, and fire suppression. These crews can range in size, with a maximum of 10 people.

Prescribed Wildland Fire Crews participate in prescribed fire and wildfire activities, which may include burn unit preparation, fire operations, mop-up, and monitoring.

Wildland Fire Modules are 7- to 10-person crews that assist in planning, fire behavior monitoring, ignition, holding, project preparation, and execution. Wildland fire modules are often assigned to fires that are being managed for multiple objectives and provide expertise in areas of fire effects monitoring, ignition, holding, line construction, and long-term planning. Wildland fire modules are required to be self-sufficient for extended periods of time and perform many of their functions in remote areas of fires or wilderness areas.

FEDERAL TOURS OF DUTY AND POSITION TYPES

The following are the different types of positions and tours of duty that are available in federal employment (US Department of Interior 2017). The states and many private organizations have similar classifications of positions.

Permanent Full-Time typically means 8 hours per day, 40 hours per week, for 52 weeks per year. Such positions include all benefits (retirement, health insurance, life insurance, annual or vacation leave, and sick leave) of permanent employment.

Permanent Part-Time includes all the benefits (retirement, health insurance, life insurance, annual or vacation leave, and sick leave) of permanent employment. Salary, vacation, and sick leave, which are normally based on a 40-hour week, are prorated based on the number of hours worked. The government contribution to health benefits premiums is also prorated and will usually mean a higher employee cost than for full-time positions at the same grade. Periods of full-time employment on a temporary basis could be scheduled depending upon workload fluctuations.

Permanent Career-Seasonal includes all benefits (retirement, health insurance, life insurance, annual and sick leave) of permanent employment, but does not provide for employment on a full-time year-round basis. In this position, the employee works at least 26 weeks, but not more than 48 weeks, in a service year. The employee works 40 hours per week when in this pay and duty status. When services are not required, such employees are placed in a nonwork, nonpay status. Salary, vacation, and sick leave earnings, which are normally based on year-round employment, are prorated according to the number of weeks worked each year. The waiting period for within-grade increases and career tenure are extended by a portion of the time spent in nonpay status.

Temporary / Seasonal is a nonpermanent position that does not provide benefits other than annual (vacation) leave or sick leave. These positions provide employment for a limited period, normally six months or less (a maximum of 1,039 hours). Employees normally work 40 hours per week. These are often called 1039 positions.

Federal General Schedule Education Requirements and Salaries

There have been several earlier references to General Schedule grade levels. The entry level requirements are given in Table 2-8 and starting annual salaries without location adjustments are given in Table 2-9. This is the federal government's method of ranking pay grades for professional positions. They use several other methods for ranking other categories of work. For example, executive positions, law enforcement officers, and skilled trades all have different schedules for pay grades. Most of the federal natural resource positions

Table 2-8   Minimum Educational Requirements for Entry Level Employment on the General Schedule (GS)

Grade	Entry Level Requirements
GS-1	None
GS-2	High school graduation or equivalent
GS-3	1 academic year above high school
GS-4	2 academic years above high school or associate's degree
GS-5	4 academic years above high school leading to a bachelor's degree, or bachelor's degree
GS-7	Bachelor's degree with Superior Academic Achievement for two-grade interval positions or 1 academic year of graduate education
GS-9	Master's degree or 2 academic years of progressively higher level graduate education
GS-11	PhD or equivalent doctoral degree or 3 academic years of progressively higher level graduate education or—for research positions only—completion of all requirements for a master's or equivalent degree
GS-12	For research positions only, completion of all requirements for a doctoral or equivalent degree

Note: After being employed with the minimum educational requirements, qualifying for jobs at the next higher grade generally requires at least one additional year of specialized experience. For example, a wildlife manager with a bachelor's degree at the GS-5 level would be required to have at least one year of specialized experience to qualify for a GS-6 position.

Professions

Table 2-9 Annual Base Salary (Step 1 only) Without Locality Pay by General Schedule Grade, 2018

General Schedule Grade	2018 Annual Base Salary (Step 1) without locality pay ($)
1	18,785
2	21,121
3	23,045
4	25,871
5	28,945
6	32,264
7	35,854
8	39,707
9	43,857
10	48,297
11	53,062
12	63,600
13	75,628
14	89,370
15	105,123

Note: Most metropolitan areas and some regions of the country have increased pay to adjust for higher costs of living in those areas.

described in this book will be for jobs on the General Schedule. States have similar systems of job classification.

Natural resources professionals in administrative positions with responsibilities beyond the scope of the General Schedule are placed on the federal Executive Schedule. Such positions are beyond the scope of the discussion in this book, but some federal natural resources professionals earn annual salaries above $140,000.

SUMMARY

The traditional professions of forester, wildlife manager, fisheries manager, and range manager are separate disciplines with similar fundamentals and processes, but each has its own advanced knowledge, skill sets, and spheres of influence. A four-year college degree with coursework specific to the discipline is needed to qualify for these professions. Natural resources professionals have wide-ranging responsibilities and are expected to exercise discretion, apply critical thinking skills, and exhibit personal accountability. Many belong to professional associations that offer opportunities for individual certification and continuing education.

STUDENT EXERCISES

1. Use the Bureau of Labor Statistics website to obtain the latest statistics on a professional interest (Forester, Forest and Conservation Technician, or Zoologist and Wildlife Manager). Examine the geographic profile and list the states and areas with the highest employment and those with the highest wages for the profession of interest.

2. Use your state's employment website to determine the starting and average wages of a job classified as *forester, forest and*

conservation technician, or *zoologist and wildlife manager.* This may take some creative searching to find the information from your state's employment data.
3. Use the USAJOBS website to search for federal natural resource jobs. Simply use the series number in the keyword search box to determine the currently available jobs in that series.

WORKS CITED

American Fisheries Society. n.d. New to Fisheries? Accessed September 14, 2017. https://fisheries.org/membership/new-to-fisheries/

Bureau of Labor Statistics. 2018a. Occupational Employment Statistics, Occupational Employment and Wages, May 2017, 19-1032: Foresters. Last modified March 30, 2018. https://www.bls.gov/oes/current/oes191032.htm.

Bureau of Labor Statistics. 2018b. Occupational Employment Statistics, Occupational Employment and Wages, May 2017, 19-1023: Zoologists and Wildlife Biologists. Last modified March 30, 2018. https://www.bls.gov/oes/current/oes191023.htm.

Bureau of Labor Statistics. 2018c. Occupational Employment Statistics, Occupational Employment and Wages, May 2017, 19-4093: Forest and Conservation Technicians. Last modified March 30, 2018. https://www.bls.gov/oes/current/oes194093.htm

Bureau of Labor Statistics. n.d. Employment Projections. Accessed April 23, 2018. https://data.bls.gov/projections/occupationProj

Fryxell, John M., Sinclair, Anthony R. E., and Caughley, Graeme. 2014. *Wildlife Ecology, Conservation, and Management*, 3rd ed. Oxford, UK: Wiley Blackwell.

Helms, John A. 1998. *The Dictionary of Forestry.* Bethesda, MD: Society of American Foresters.

Idaho OnePlan. n.d. Range Management. Accessed September 14, 2017. http://oneplan.org/Range/index.asp.

Office of Personnel Management. 2005. *Professional Work in the Natural Resources Management and Biological Sciences Group, 0400.* Accessed August 20, 2017. https://www.opm.gov/policy-data-oversight/classification-qualifications/classifying-general-schedule-positions/standards/0400/gs0400p.pdf

Office of Personnel Management. 2017a. Classification and Qualifications, General Schedule Qualification Standards. List by Occupational Standards. Accessed August 20, 2017. https://www.opm.gov/policy-data-oversight/classification-qualifications/general-schedule-qualification-standards/#0400-ndx=&url=List-by-Occupational-Series.

Office of Personnel Management. 2017b. Classification and Qualifications, General Schedule Qualification Standards. Group Standards. Accessed August 20, 2017. https://www.opm.gov/policy-data-oversight/classification-qualifications/general-schedule-qualification-standards/#0400-ndx=&url=Group-Standards

Society of American Foresters. 2017. State Licensure. Accessed September 10, 2017. https://www.eforester.org/Main/Certification_Education/Licensure/Main/Certification/Licensure.aspx?hkey=a0c534f7-c6da-4749-ad5f-e37da299b97c.

State of Oregon Employment Department. n.d. *Occupational Profiles.* Accessed August 30, 2017. https://www.qualityinfo.org

The Wildlife Society. n.d. *Careers.* Accessed September 14, 2017. http://wildlife.org/next-generation/career-development/careers/.

US Department of the Interior. 2017. *Wildland Fire Jobs.* Last updated November 28, 2017. https://www.firejobs.doi.gov/index.php?action=crews

Wray, C. 1971. Forestry Technician: Noncom of the woods. *Journal of Forestry* 68(8): 498–500.

ADDITIONAL RESOURCES

Society of American Foresters Career Center
http://careercenter.eforester.org/

The Wildlife Society Careers
http://wildlife.org/next-generation/career-development/careers/

3

History

Chapter Outline
Developing the Idea of Conservation, Civil War to 1890
Conservation and Wise Use, 1891 to 1930
Management Period, 1930 to 1960
Concern about Quality of Life, late 1950s to 1970
Global Concerns, 1970s to Present Day

The movement to conserve natural resources is less than 150 years old in the United States, and the term *conservation* itself has only been in use for about 100 years. During the first 50 years of the conservation movement, the idea that natural resources could and should be managed in a logical manner gradually gained acceptance. It took another 40 years or so to develop the science, organizations, professions, and educational institutions that would support the concept of conservation. Over the past 50 years the movement has evolved to include increasing public input and social influence. The rich history of this 150-year span includes two world wars, a worldwide depression, a "cold" war, and the rise of the United States from an isolated, agricultural nation to an industrialized, urbanized world power. The history of the conservation movement parallels the history of the nation and its generations of people. In other words, the development of the conservation movement and its culture of agencies, companies, and professions cannot be separated from the broader history and culture of the United States and, increasingly, the entire world.

An understanding of the history of the conservation movement will help provide context to our discussion of the development of the forestry, wildlife management, fisheries, and range management professions. The evolution of the conservation movement has not been evenly paced. Changes came about very slowly in the late 1800s, and more rapidly during the 1930s and 1960s. Regardless of their speed, the developments were driven by a complex interplay of scientific advances, education, public perception, politics, and individual personalities. This process continues today against a backdrop of increasing populations, competition for resources, and a shift in focus from local issues to global concerns.

The forests, woodlands, grasslands, and waters of the nation represent tangible wealth from the extraction of their resources and intangible wealth from an appreciation of their beauty. People are passionate about our natural resources, but have different views on how they should be used and managed. These differences have led to spirited discussions, arguments, and political battles. Understanding these different perspectives and placing the current situation into a broader historical context reminds students and professionals alike that people who work in natural resources operate within a changing environment within a dynamic society that values its natural resource for different purposes. There will be ongoing tension over resource uses, balancing private property rights with the public good, statutory law with regulatory law, and state powers with federal authority. Natural resources professionals must constantly contemplate the role of humans within the natural environment and the best way to manage and allocate natural resources for our current and future benefit—within the context of a democratic society. A historic perspective is useful for this task.

We can begin to develop this historic perspective by considering the following questions: How did the traditional professions come to exist? What does it mean to manage natural resources within the governmental and societal systems that exist in the United States? This book does not address natural resource policy making or policy issues. Rather, it reinforces the fact that such policy decisions are made in the context of management objectives that are influenced by history, and therefore provides a brief overview of the development of resource conservation in the United States.

The history of the US conservation movement begins by acknowledging that the land area of the United States was occupied and influenced by native peoples before the influence of other cultures. In some regions, the impact of Native Americans was quite significant. Unfortunately, we do not have extensive records of the vegetation type and structure or the fish and wildlife populations of the United States before European influence. What we do know is that the European migration and subsequent settlement had a profound effect on the American landscape.

The following history is not exhaustive or detailed; instead, it presents the highlights and places the development of the traditional natural resource professions in a historical context. Hendee et al. (2012)

described four periods of conservation in the United States. The following discussion modifies this into five somewhat overlapping periods of conservation. It also discusses the development of the science of ecology and the individual natural resource professions. Rather than clutter the text with multiple citations from these sources, the bulk of the history of conservation can be credited to Hendee et al. (2012); and the history of ecology credited to Worster (1994) and encyclopedia.com (2017). Other sources will be cited in the discussion.

DEVELOPING THE IDEA OF CONSERVATION, CIVIL WAR TO 1890

The conservation movement had its beginnings in the mid-1800s with a call for change to address wasteful practices. Prior to this, the period of European migration and settlement witnessed extraction of natural resources with little or no consideration of sustainability. There were local shortages of natural resources, and therefore some local attempts at regulation. But the prevailing view was that the supply of natural resources was inexhaustible and that new sources could always be found on the country's frontier. The realization that natural resources were, in fact, limited—particularly the resources of wood, water, and wildlife—was a long, slow process of growing awareness.

Industrialization in the United States surged after the Civil War (1861–1865). Some historians call this period a second industrial revolution. Before that time, much of the country's economy depended on wood resources, not just as building material but also as the dominant source of energy. Coal did not overtake wood as an energy source until the mid-1880s, and the use of petroleum for energy did not surpass the use of wood until about 1910 (US Energy Information Administration 2013). Wood-hulled ships were prevalent until the late 1800s. The nation's use of wood in 1850 was about 5 billion board feet (215 board feet per capita) and rose to 35 billion board feet in 1900 (about 407 board feet per capita) (US Bureau of the Census 1908). A **board foot** is defined as a piece of wood one inch thick and one foot square, or 144 cubic inches of wood. It is against this backdrop of increased mechanization and industrialization, post-Civil War reconstruction of the South, and increasing demand for forest resources that the idea that natural resources should be protected and conserved began to take root.

Four Americans were particularly influential early in this period. Ralph Waldo Emerson (1803–1882) and Henry David Thoreau (1817–1862) were leaders in the transcendental movement, a philosophical and literary movement that advocated the inherent goodness of nature and of people who were self-reliant and independent thinkers rather than conformists to society and its institutions. Both wrote essays and books about the importance of nature; Emerson's best-known work is his essay *Nature* (1836), while Thoreau's *Walden* or *Life in the Woods* (1854) has become a classic. George Perkins Marsh (1801–1882) published an influential book entitled *Man and Nature* in 1864, and George Bird Grinnell (1849–1938), editor of *Field and Stream*, advocated for the preservation of the American bison, helped found the Boone and Crocket Club, and founded a precursor to the Audubon Society. These four writers and others promoted the idea that nature had an intrinsic value and was an important aspect of American culture.

Also during this period, scientists were advancing a better understanding of natural biological systems. Charles Darwin published *Origin of Species* in 1859, which profoundly influenced biological thought and continues to do so. It is claimed that Darwin is the single most important person in the history of ecology (Worster 1994), and that Darwin's work caused America's Ivy League universities to pay attention to the natural sciences which, in turn, produced graduates with an interest in the natural sciences—an interest that eventually led to setting aside public lands as reserves (Brinkley 2009). It was during this period that Ernst Haeckel, a German biologist, used the term *ecology* for the first time in 1866. The stage was set for the emergence of a new, biologically dependent, interdisciplinary science of ecology. This new science would eventually change the way people thought about natural systems and their importance and value to society.

The visual arts also played a role in influencing attitudes toward nature in the United States. The Hudson River School—an art movement of the mid-1800s—was embodied by a group of popular American landscape artists. Their scenic and often moralistic paintings initially captured the natural beauty of the Hudson River Valley, the Catskills, and the Adirondacks, and later expanded to images of the American West—including Yellowstone and Yosemite. The imagery struck a chord with the American public and served as propaganda for the early conservation movement. The works produced by these artists introduced images of the American West to those in the East.

As a greater appreciation for nature was being advanced by writers and artists, government agencies slowly began to take action. Preservation of unique land areas began in 1864 when President Abraham Lincoln signed the Yosemite Grant, which set aside from development some of the area that was to later become Yosemite National Park. The federal government later ceded the area to California as a state park. In the meantime, the Yellowstone National Park Act of 1872 designated Yellowstone as a park for "a public pleasuring ground" and for preservation. Because of the difficulties of controlling the area and the Department of Interior's lack of authority to police the park area, control of Yellowstone National Park was turned over to the US Army in 1886. In 1890, Yosemite was declared a national park by the federal government and most of the park area earlier ceded to California reverted to federal control and was turned over to the US Army for protection. There was an increasing awareness that government action needed to go beyond just reserving the land and extend to physical protection. However, the general attitudes of the time did not favor the widespread federal protection of resources.

The public began to recognize its role in conserving resources with the founding of the American Fisheries Society in 1870, the American Forestry Association in 1875, and the American Forestry Congress in 1882 (which soon merged with the American Forestry Association). Hunting and fishing clubs were being formed, and numerous periodicals geared to sportsmen further influenced public perception of the need for conservation (Brown 2007). About this time, John Muir emerged as

President Theodore Roosevelt (left) and influential preservationist John Muir (right), founder of the Sierra Club, on Glacier Point in Yosemite National Park in 1903. (US Library of Congress)

an articulate voice for preservation. He began publishing essays and books on conservation issues in 1874, helped convince the federal government to protect Yosemite as a national park in 1890, and co-founded the Sierra Club in 1892. Theodore Roosevelt, a self-proclaimed fan of Charles Darwin who later became president, began to publish books on hunting and western ranch life during the 1880s. He was a co-founder of the Boone and Crocket Club in 1887. The club worked to end market hunting (hunting for the commercial sale of meat, hides, feathers, or bones) and promoted wildlife conservation and the concept of free-chase hunting in which the big game is wild and free roaming. These civilian organizations were beginning to sway both public opinion and federal natural resource policy.

Meanwhile, the federal government was forming agencies to study and address conservation issues. The United States Commission of Fish and Fisheries was formed in 1871 to address the issue of declining marine and freshwater fisheries. The Division of Forestry within the USDA was established in 1881. Its first chief was Dr. Franklin Hough, a physician, who wrote the country's first forestry book, *Elements of Forestry*, published in 1882. The task of the Division of Forestry was initially to gather information and later to conduct experiments (Forest History Society 2017). In 1886, Bernhard Fernow, a German-educated forester, became chief of the division. These government agencies did not yet control the resources or land area; their role was to gather the underlying information needed to compel more intense regulation of the resource.

CONSERVATION AND WISE USE, 1891 TO 1930

The stage had been set for the US government to transform public awareness and acceptance of resource conservation into policy. This effort began in earnest in 1891 with the Forest Reserve Act. This act largely stopped the disposal of federal land and authorized the president to preserve timberland as "forest reserves." Before this time the federal policy was to dispose of land to states, individuals, or corporate enterprises. The Forest Reserve Act marked the closing of the American frontier, at least for the continental portion of the United States, and the end of "free" land. This marked the beginning of the second period of conservation in the United States during which there was a shift from the custodial holding of land and resources to a period of professional management.

At the outset of this wise-use period, President Harrison used the authority of the Forest Reserve Act to create 15 reserves with an area greater than 13 million acres by 1893. President Cleveland added another 5 million acres to the reserves, and then stopped—realizing that there was no plan for their management. A National Forest Commission was formed in 1896 and although the commission was fraught with disagreements, it recommended the establishment of two national parks, Mt. Rainier and Grand Canyon, and more forest reserves. President Cleveland, heeding the commission's recommendations, added 13 new reserves with over 21 million acres just before leaving office in 1897, despite there being no plan for the land's management. Later that year the Forest Reserve Organic Act was passed, which continued to allow the president to create forest reserves for producing timber and protecting water and authorized the Department of Interior to manage the land in the reserves. President McKinley added another 7 million acres to the reserves. By the turn of the century there were some 46 million acres of federal forest reserves.

Theodore Roosevelt, who served as president from 1901 to 1909, became known as the "Conservation President" and the namesake for the Teddy Bear. Teddy Roosevelt was a Harvard educated naturalist turned politician. Roosevelt had a close relationship with Gifford Pinchot, a Yale graduate with European training in forestry. Both men were from elite, wealthy families and raised in New York City. In 1908 Roosevelt sponsored a Governors' Conference on the Conservation of Natural Resources. This conference marked the widespread establishment of the conservation movement in the United States (Moffitt 2001), and many states broadened their interest in and regulation of natural resources because of the conference. During his presidency Roosevelt created five new national parks, including Crater Lake and Mesa Verde National Parks; signed the 1906 Antiquities Act that authorizes the president to protect landmarks, structures, and objects of historic or scientific interest by designating them as "national monuments," and then created 18 national monuments, including the Grand Canyon (later a park); established 51 bird preserves and four wildlife preserves; and placed 230 million acres of land under increased federal protection.

Gifford Pinchot had been appointed chief of the USDA Division of Forestry when Fernow quit in 1898, and he presided over several administrative changes. In 1901 the Division of Forestry became the Bureau of Forestry and Pinchot continued as chief forester. The problem of having the Forest Reserves in the Department of Interior and the foresters in the Department of Agriculture's Bureau of Forestry was remedied by the Transfer Act of 1905, through which some 63 million acres of Forest Reserves were placed under the control of the Department of Agriculture. The act also transferred 500 people from the Department of the Interior to the Department of Agriculture. The Bureau of Forestry was renamed the Forest Service two months later and, in 1907, the Name Change Act renamed the Forest Reserves as National Forests.

Reenactors portray Teddy Roosevelt, the "Conservation President," and Gifford Pinchot, first Chief of the Forest Service. (USFS Mt. Hood National Forest, Flickr)

Active outside the federal government as well, Pinchot and his parents helped found the Yale School of Forestry in 1900. He was also instrumental in establishing the Society of American Foresters in 1900. Pinchot remained as Chief of the Forest Service after Roosevelt left office in 1909, but was fired by President Taft early in 1910 after a confrontation with Taft's Secretary of the Interior over land management policies of the Department of Interior.

Pinchot remained active in the conservation movement, leading one faction of the movement through a political battle that became known as the Hetch Hetchy controversy. In 1908, the growing city of San Francisco proposed building a dam in the Hetch Hetchy Valley to provide a steady water supply. The Hetch Hetchy Valley was within Yosemite National Park and protected by the federal government. National opinion was divided between those favoring the wise use of resources to benefit people and giving San Francisco the right to dam the valley, and the preservationists who wanted to protect the valley from development. John Muir represented the proponents of preservation and Gifford Pinchot represented the proponents of wise use. In 1913 Congress decided to build the dam and flood the valley. Although the preservationists were unsuccessful in this battle, the increased public awareness generated by the political effort helped lead to the creation of the National Park Service in 1916. The preservationists and wise-use factions of the conservation movement never fully reconciled into one unified voice. The Hetch Hetchy controversy has reemerged, with recent proposals to remove the dam and begin restoring the Hetch Hetchy Valley.

Major federal acts during this second phase of the conservation movement included:

- The Lacy Act (1900), which outlawed the movement of illegally taken game across state lines.
- The Weeks Act (1911), in reaction to the fires of 1910, this act authorized cooperation in fire control between states and federal resources and authorized the purchase of cutover lands in the eastern United States for the protection of watersheds.
- The National Park Service (1916), created within the Department of Interior, as a direct response to the decision to build the dam in the Hetch Hetchy Valley.
- Migratory Bird Treaty Act (1918) set standards for the hunting and management of birds migrating between the United States, Mexico, and Canada.
- Migratory Bird Conservation Act (1929) authorized the purchase of land for waterfowl refuges.

The period of the 1890s to the 1930s also witnessed the development of the science of ecology. During the 1880s and 1890s, Stephan Alfred Forbes influenced the development of aquatic ecology at the University of Illinois and the Illinois State Laboratory of Natural History; Edward Birge advanced the development of limnology through his work at the University of Wisconsin; Henry Cowles, a University of Chicago professor, influenced terrestrial ecology through his studies of the Indiana Dunes on the shore of Lake Michigan; and Fredric Clements studied plant communities on the prairies of Nebraska. Cowles and Clements defined the concept of the ecological succession of communities. They adapted the term "succession" from the ideas of Henry David Thoreau. Clements' concept of a stable state, or climax community, which was highly influenced by the local climate, would dominate ecological thought until the 1950s. Charles Elton published a book on animal ecology in 1927 in which he promoted the concepts of food cycles or chains, the niche, and the pyramid of numbers. In the 1920s and 1930s ecology became more quantitative and mathematical. It was maturing as a science and increasingly influenced the management of natural resources.

This period also marked the beginning of formal natural resource education in the United States. Before this time no American universities taught forestry or fisheries as distinct programs. In 1898 Dr. Carl Schenk, a German forester, started the Biltmore School, an intense, applied bachelor's degree program. It was run at the Biltmore Estate outside Asheville, North Carolina until 1909, and then operated as a travelling school until 1913. The typical pattern for instruction was lectures in the morning, with afternoons devoted to practical lessons in the forest. Although this program was short lived, it marked the beginning of practical forestry education in the United States, and its pattern of morning lectures and afternoon field labs is still being used in many natural resource programs.

Other natural resource education programs were also initiated during this time span. Bernhard Fernow was instrumental in starting the forestry college at Cornell University in 1898. This college closed in 1903, but forestry courses continued to be taught at Cornell. The Yale School of Forestry opened in 1900 and held summer camps at Grey Towers, the Pinchot estate in Milford, Pennsylvania. The Yale School of Forestry still operates. Other universities soon developed forestry programs, and in 1919 the first full university program in fisheries was established at the University of Washington.

MANAGEMENT PERIOD, 1930 TO 1960

This period is defined by three profound events in US history: the Great Depression, World War II, and the beginning of the baby boom, all of which impacted the conservation movement as it was beginning to mature. Franklin Roosevelt, who served as president from 1933 to 1945, was a strong supporter of conservation issues. In 1934 he established the Civilian Conservation Corps (CCC), a federally funded organization that put thousands of Americans to work during the Great Depression on projects with environmental benefits. The CCC employed three million young men in such projects as tree planting, erosion and flood control, road and trail building, firefighting, and construction, most of which was carried out on Forest Service land. These projects, along with other federal programs, such as the Works Progress Administration (WPA), increased recreational access to federal lands. The CCC was dismantled in 1942 when Congress ceased its funding due to the demands of US involvement in World War II. The war effort was the major political focus of the country and conservation issues were largely dormant during the war.

Notable federal actions for funding wildlife and fisheries during this period included the 1934 Migratory Bird Hunting Stamp Act, which required the purchase of a federal stamp for hunting migratory waterfowl; the 1937 Pittman-Robertson Act, which taxed sporting arms and ammunition to provide funding for wildlife research and preserves; and the 1950 Dingell-Johnson Act, which taxed sales of fishing equipment to provide funding for fish restoration and management.

One notable international development with regard to the regulation of natural resources was the formation of the International Whaling Commission in 1946 with 15 member countries. The commission was formed for conservation, management, and research of whales and whaling.

Ecological science and the conservation professions also developed during this time frame. In 1935 Sir Arthur Tansley coined the term "ecosystem," the concept of a community of organisms and their nonliving environment. Eugene Odum further developed the ecosystem concept in his formative textbook *Fundamentals of Ecology*, written with his brother Howard in 1953. This book became the primary ecology text in US universities for the next decade.

The discipline of wildlife management was established in the 1930s (Brown 2007). Aldo Leopold, a Yale-trained forester, took a leading role in the formation of the wildlife management profession. In 1933 Leopold was appointed "professor of game management" at the University of Wisconsin, the first such appointment of this type in the United States. In 1933 his book *Game Management* was published, and in 1937 he helped found The Wildlife Society. His best-known contribution to conservation was his 1949 book, *A Sand County Almanac*, which was published posthumously and did not gain widespread popularity until the 1970s. In it, Leopold wrote about a "land ethic," a philosophic concept that emphasized the relationship existing between people and the land they inhabit. The discipline of fisheries science emerged in the 1950s (Royce 1985), and the Society for Range Management was founded in 1948. The conservation movement now had four distinct disciplines: forestry, wildlife biology, fish biology, and range management—all being taught at the university level and each with a professional society.

The end of this so-called management period witnessed economic expansion, rural flight, population boom, and suburban development. The demand for lumber increased with suburban growth and economic expansion and so too the demand for foresters. Because timber production during World War II was primarily from private land and those stocks were depleted, the national forests began to ramp up timber harvests to fulfill the demand for lumber. The Interstate Highway System, championed by President Eisenhower, was funded to connect the principal metropolitan areas, cities, and industrial centers, and to serve the national defense. It also facilitated faster and safer recreational travel by the American public. Enhanced transportation networks combined with other factors—higher incomes, suburban living, and military and CCC veterans who identified with the outdoors—to significantly boost recreational use of wildland parks and forests. Both the national parks and the national forests responded by increasing their capacity to accommodate recreational visitors. Enrollments in natural resource programs increased as veterans took advantage of the GI Bill to attend college.

CONCERN ABOUT QUALITY OF LIFE, LATE 1950s TO 1970

This period started with the can-do attitude of the "greatest generation"—Americans who grew up during the Great Depression and fought in World War II—and ended with the protests of the early baby boomers. It was a period of social unrest in the United States and significant change in the conservation movement. Rachel Carson, a marine biologist, warned of the effects of pesticides in her book, *Silent Spring*, first published as a series in *The New Yorker* in 1962. It is a significant work that challenged the use of the pesticide DDT, questioned society's reliance on technological advances, raised the issue of quality of life, and helped launch the environmental movement. Carson was one of many voices during this period that called for preservation of natural resources and opposed wise-use management.

Policy changes affecting conservation agencies were rapid and sustained throughout this period. Some of these included the following:

- Multiple-Use, Sustained-Yield Act (1960) directed the Forest Service to manage for multiple uses and with a sustained yield of harvestable resources.
- McIntire-Stennis Cooperation Research Act (1962) funded forestry and other wildland research at universities and experiment stations.
- National Outdoor Recreation Act (1963) created a bureau to study and plan outdoor recreation opportunities. It eventually was folded into the National Park Service.
- Clean Air Act (1963) was the first federal legislation aimed at air pollution control.
- Land and Water Conservation Act (1964) created a fund using offshore oil-lease monies to fund the acquisition of land for federal conservation programs and grants for states to do the same.
- Wilderness Act (1964) created the National Wilderness Preservation System.
- Water Quality Act (1965) set water quality standards and addressed water pollution issues.
- National Historic Preservation Act (1966) established a process to protect historic and archaeological sites.
- National Trails Act (1968) proclaimed the Appalachian and Pacific Crest trails as national scenic trails and allowed for the creation of new scenic and historic trails.

- Wild and Scenic Rivers Act (1968) protected selected rivers for their wild or scenic value.
- National Environmental Policy Act (1969) required agencies to examine the environmental consequences for most federal projects conducted by executive agencies.

These were sweeping changes, particularly those mandated by the National Environmental Policy Act, which went into effect on January 1, 1970. States soon adopted similar laws, such as the 1970 California Environmental Quality Act, requiring an examination of environmental consequences for any state government or state-funded action within California. These laws dramatically altered the planning processes of the affected federal and state agencies and changed the culture of the agencies.

Another significant change in federal agency culture was brought about by the creation of legal Wilderness Areas in 1964. By legal mandate, these areas were to be managed differently than adjacent land areas and this has been an ongoing challenge for federal land managers. One challenge is accomplishing work without the use of vehicles or motorized equipment. Except in emergencies, motorized equipment like chainsaws and motorized vehicles are not allowed in Wilderness Areas. This includes the landing of aircraft. All non-emergency work is to be accomplished using hand tools even if the cost is greater than using motorized equipment. Another example of the challenges of wilderness management is the ongoing debate over the interpretation of "mechanical transport" as written in the Wilderness Act. Initially, bicycles and other wheeled devices, such as strollers, were allowed in Wilderness Areas; they were later banned when mechanical transport was interpreted more strictly. Currently, there is mounting pressure to allow bicycles and other wheeled devices in Wilderness Areas, including a congressional bill to amend the Wilderness Act. Wheelchairs, however, are allowed in Wilderness Areas because the Americans with Disabilities Act (1990) specifically provided for wheelchair access but does not require the improvement of Wilderness Area trails to accommodate wheelchairs. So, wheels are allowed on wheelchairs, but not on other devices, except in emergencies.

One cannot overemphasize the change in natural resource agency culture due to the laws and policy shifts of the 1960s and 1970s. Natural resource agencies were faced with changes in their planning processes and documentation, as well as a broader charge to provide clean water and air.

GLOBAL CONCERNS, 1970S TO PRESENT DAY

This period is characterized by a growing concern for the broader impacts of human activities on regions, continents, and the whole planet. One of the first significant events was passage of the Environmental Quality Improvement Act in April 1970, which worked in conjunction with the 1969 National Environmental Policy Act. The Improvement Act created an Office of Environmental Quality and gave it responsibility for ensuring that federal departments and agencies follow federal environmental policy and law. Soon after, the first Earth Day was celebrated in April of 1970. Earth Day observances have been held ever since and are a combination of celebration, educational forum, and protest centering on environmental protection.

Starting in the 1970s, environmental concerns evolved from local and regional concerns to global concerns. Examples of this progression include the acid rain controversy, concern over tropical deforestation, the save the whales movement, the spotted owl controversy and the resulting Northwest Forest Plan, concern over ozone depletion and the ozone hole in the Antarctic, and the continued scientific discussions and political arguments over global climate change (initially global warming). It became what some call the period of the "ologists" (biologists, ecologists, archeologists, hydrologists), during which agencies and companies scrambled to hire these specialists to address the increasingly complex issues surrounding new environmental policies and resulting regulations.

Policies and laws enacted during this period include the following:

- Marine Mammal Protection Act (1972), which prohibits taking marine mammals in US waters and by US citizens on the high seas and the importation of marine mammals and their parts.
- Endangered Species Act (1973), based on earlier acts, resulted in comprehensive protection for plants and animals of all types, not just game animals.
- Fish Conservation and Management Act (1976), which restricted foreign fishing in US waters.
- National Forest Management Act (1976), which required public participation in the national forest planning process and in any significant changes to the management of the national forests. This act, in part, resulted from controversy over clearcutting in the Monongahela National Forest and clearcutting and terracing in the Bitterroot National Forest. The public objected to these management practices and to not being informed and allowed to comment prior to their implementation.

- Clean Water Act (1977), amended and strengthened the 1972 Clean Water Act.
- National Indian Forest Resource Management Act (1990), provided a framework for forest management for the Bureau of Indian Affairs.
- National Parks Omnibus Management Act (1998), provided an overhaul of park management and operations.
- Healthy Forest Restoration Act (2003), a controversial act aimed at streamlining the process of reducing hazardous fuel levels in fire-prone areas.
- Omnibus Public Land Management Act (2009), created a National Landscape Conservation System within the Bureau of Land Management, support for collaborative groups, and added two million acres to Wilderness Areas.
- Federal Land Assistance, Management, and Enhancement Act, FLAME (2009), created funding for suppression costs of large wildland fires and mandated the creation of a cohesive wildfire management strategy.

Forestry professionals collaborate on a restoration project on the Sawtooth National Forest. Natural resources professionals must be good communicators and seek public input on projects. (National Forest Foundation, Flickr)

Changes in natural resource management were also occurring on the state and global levels as well. During the 1970s, Oregon, Washington, and California enacted comprehensive forest practice acts placing wide-ranging regulations on practices on both private and state-owned forests. These three states and others continue to update their forest practice policies and regulations. Global environmental concerns also compelled changes in policy and regulations. The concern over ozone loss in the upper atmosphere resulted in the Vienna Convention for the Protection of the Ozone Layer in 1985 and the subsequent Montreal Protocols in 1987 that yielded international regulation of chlorofluorocarbons that were being used as refrigerants and aerosol propellants. The Montreal Protocols were ratified by all the members the United Nations. Concerns over the increasing concentrations of greenhouse gases led to the United Nations Framework Convention on Climate Change in 1992, Kyoto Protocol in 1997, and the Paris Climate Agreement of 2015. Although controversial, these agreements indicate the ongoing international concern over global environmental issues.

In addition to forestry, range, wildlife, and fisheries programs, universities and colleges began offering broader degrees in natural resources, environmental sciences, environmental studies, and sustainability. Eventually hundreds of universities would offer these new programs. Enrollments in natural resources programs increased until the early 1980s, then declined until peaking again in the mid-1990s. More details on enrollments are provided in Chapter 5.

SUMMARY

The conservation movement in the United States has been dynamic. It fostered the formation of conservation organizations and redefined the role of the federal and state governments—from a policy of transferring land to states and private entities, to custodial care of land, to active management of land and natural resources, to the creation of agencies. The early conservation movement was quite unified. Later, a rift developed between those who advocated wise use of resources and the preservationists, as evidenced by the Hetch Hetchy controversy. The two factions never fully reconciled but often worked in parallel through the 1950s. There was a resurgence in the preservation movement with the advent of the environmental protection causes of the 1960s.

As the ideals of conservation and preservation gained traction, so too did efforts to offer educational programs and professional training in forestry, wildlife management, fisheries management, and range management. As these professions developed they have redefined themselves to align with societal concerns and preferences with regard to the natural resources. This process will no doubt continue as larger, global issues are confronted.

STUDENT EXERCISES

1. View the video, *The Greatest Good* (A Forest Service Centennial Film), and outline the history of the Forest Service, emphasizing the changes that occurred in its objectives and the country's political environment from 1905 to 2005.
2. View the video, *The National Parks, A Centennial History*, and outline the history of the national parks, emphasizing the political events that led to their creation and the creation of the National Park Service.
3. Read *A Sand County Almanac* and explain the concept of "land ethic." Describe how it might be applied by a public land management agency.
4. Choose one of the four traditional professions—forestry, wildlife biology, fish biology, or range management. Write a brief history from its origin to the present and then speculate on the future of this profession.

WORKS CITED

Brinkley, D. 2009. *The Wilderness Warrior: Theodore Roosevelt and the Crusade for America*. New York: Harper Collins.

Brown, R. D. 2007. The history of wildlife conservation and research in the United States—with implications for the future. In H. Li (Ed.), *Proceedings of the Taiwan Wildlife Association 2007* (pp. 1–30). Taipei: Taiwan National University.

encyclopedia.com. 2017. Ecology, History of. Accessed September 17, 2017. http://www.encyclopedia.com/plants-and-animals/botany/botany-general/history-ecology

Forest History Society. 2017. US Forest Service History: Agency Organization. Accessed February 12, 2018. https://foresthistory.org/research-explore/us-forest-service-history/policy-and-law/agency-organization/

Hendee, J. C., Dawson, C. P., and Sharpe, W. F. 2012. *Introduction to Forests and Renewable Resources*, 8th ed. Long Grove, IL: Waveland Press.

Moffitt, C. 2001. *Reflections: A Photographic History of Fisheries and the American Fisheries Society in North America*. Bethesda, MD: American Fisheries Society.

Royce, W. F. 1985. The Historical Development of Fisheries Science and Management. Taken from a lecture given at the Fisheries Centennial Celebration. NOAA website. Accessed September 21, 2017. https://www.nefsc.noaa.gov/history/stories/fsh_sci_history1.html

US Bureau of the Census, Department of Commerce and Labor. 1908. *The Lumber Cut of the United States 1907: Compiled in Cooperation with the Department of Agriculture: Forest Service, Gifford Pinchot, Forester*. Washington, DC: Government Printing Office.

US Energy Information Administration. 2013. Today in energy: Energy sources have changed throughout the history of the United States. July 3, 2013. Accessed February 12, 2018. https://www.eia.gov/todayinenergy/detail.php?id=11951

Worster, D. 1994. *Nature's Economy: A History of Ecological Ideas*, 2nd ed. New York: Cambridge University Press.

ADDITIONAL RESOURCES

Bonesteel, P. 2015. *America's First Forest: Carl Schenck and the Asheville Experiment*. Durham, NC: Forest History Society.

Cronon, W. (Ed.). 1996. *Uncommon Ground: Rethinking the Human Place in Nature*. New York: W. W. Norton.

Duncan, D., and Burns, K. 2009. The National Parks: America's Best Idea. PBS Videos.

Duncan, D., and Burns, K. 2009. *The National Parks: America's Best Idea—An Illustrated History*. New York: Knopf.

Fox, S. 1985. *The American Conservation Movement, John Muir and His Legacy*. Madison: University of Wisconsin Press.

Green Fire: Aldo Leopold and a Land Ethic for Our Time. 2011. DVD. Aldo Leopold Foundation, US Forest Service, and Center for Humans and Nature.

Leopold, A. 1949. *A Sand County Almanac, and Sketches Here and There*. New York: Oxford University Press.

Lewis, J. G. 2005. *The Forest Service and The Greatest Good, A Centennial History*. Durham, NC: Forest History Society.

Miller, C. 2001. *Gifford Pinchot and the Making of Modern Environmentalism*. Washington, DC: Island Press.

Miller, C., and Staebler, R. 2004. *The Greatest Good, 100 Years of Forestry in America*, 2nd ed. Bethesda, MD: The Society of American Foresters.

Nash, R. F. 2001. *Wilderness and the American Mind*, 4th ed. New Haven, CT: Yale University Press.

Rutkow, E. 2012. *American Canopy: Trees, Forests, and the Making of a Nation*. New York: Scribner.

US Department of Agriculture, Forest Service. 2005. The Greatest Good. A Forest Service Centennial Film. Three-disk DVD set.

4

Agencies

Chapter Outline
United States Federal Government
    Department of Interior
    Department of Agriculture
    Department of Defense
    Department of Commerce
    Department of Energy
    Environmental Protection Agency
    Quasi-Government Agencies
Tribal Governments and Alaska Native Corporations
State and Local Governments
Other Employers of Natural Resources Professionals
    Nongovernment Organizations
    Educational Organizations
    Private Landowners and Managers

For whom do natural resources professionals work? Foresters, wildlife managers, fisheries managers, range managers, and other natural resource specialists work at the federal, state, and local government level; they work for quasi-government and nongovernmental organizations; they work for large and small private companies; and they work for international organizations. Additionally, they can work for themselves—serving clients who hire them for specific tasks. A surprising number of natural resources professionals are involved in teaching; in May 2017 there were about 2,000 college instructors teaching forestry, conservation science, and forest technology (Bureau of Labor Statistics 2018a).

Agencies

Each agency and company has its own culture, and the culture will vary even among units within the organization. For example, a small field office far removed from the headquarters of a state wildlife agency may have a far less formal atmosphere than the main office in the state's capital. The culture and working atmosphere may change seasonally as well. Natural resources jobs are seasonal, field-dependent positions with summer and winter project seasons, wildland fire seasons, hunting and fishing seasons, and recreational seasons that all influence the atmosphere of an agency and its offices. Toward the end of summer when deadlines to complete projects are near and wildland firefighting is stripping resources from other projects, the working environment can become stressed.

As in previous chapters, the term *wildlife management* will be used for both wildlife biology and wildlife management and the term *fisheries management* will be used for both fisheries science and fisheries management.

State governments, as a group, employ the most foresters, wildlife managers, fisheries managers, and zoologists (see Tables 4-1 and 4-2, reproduced from Tables 2-1 and 2-4). As mentioned in Chapter 2, the Bureau of Labor Statistics lumps wildlife, fisheries, and zoology professionals into one employment category so it is difficult to separate out exact differences in the employment of these three professions. Federal governments employ the second highest number of foresters, followed closely by local governments. For wildlife and fisheries managers the

Table 4-1 Employment of Foresters in the United States, 2017*

Foresters	Number of Jobs	Mean Annual Salary ($)
State governments	3,010	57,790
Federal executive branch	1,270	65,020
Local governments	1,100	61,520
Sawmills and wood preservation industry	580	62,500
Logging industry	580	64,470
Electric power generation, transmission and distribution	300	75,310
Management of companies and enterprises	200	70,630
Pulp, paper, and paperboard mills	160	68,980
Veneer, plywood, and engineered wood product manufacturing	100	68,930
Other	1,000	
Total	**8,300**	**61,710**

Note: Does not include self-employed individuals or owners/partners in unincorporated firms. (*Source:* Bureau of Labor Statistics 2018a.)

Table 4-2 Employment of Wildlife Managers, Fisheries Managers, and Zoologists in the United States, 2017*

Wildlife, Fisheries, and Zoology	Number of Jobs	Mean Annual Salary ($)
State government	6,650	57,190
Federal executive branch	4,120	82,490
Management, scientific, and technical consulting services	1,410	75,840
Local government	1,320	64,370
Colleges, universities, and professional schools	1,270	61,860
Scientific research and development services	900	63,650
Social advocacy organizations	550	62,570
Architectural, engineering, and related services	430	64,090
Management of companies and enterprises	110	80,240
Other	950	
Total	**17,710**	**66,250**

* *Note:* Does not include self-employed individuals.
(*Source:* Bureau of Labor Statistics 2018b.)

federal government is the second largest employer, employing significantly more wildlife and fishery personnel than local governments.

UNITED STATES FEDERAL GOVERNMENT

There were more than 44,000 natural resources professional and technician positions in the federal government as of September 2017 (Table 4-3). Of these, 13,538 are in general natural resources management and biological sciences (natural resource scientists). The federal government employed 15,889 forestry technicians, 5,167 park rangers, 2,281 wildlife managers, 2,243 fisheries managers, 1,851 range technicians, 1,775 foresters, 902 range managers, 583 wildlife refuge managers, and 367 fish and wildlife administrators. These September 2017 numbers (as compared to March 2017) reflect the seasonal nature of natural resource employment, particularly with regard to forestry technicians, range technicians, and park rangers.

About 43 percent of all these positions were in the Forest Service. Within the Forest Service, more than half of their natural resource positions are forestry technicians. The Forest Service employs 1,326 foresters, about 75 percent of all federal forester positions. The Forest Service also employs the highest number of wildlife managers (598),

Table 4-3 Number of federal positions in select 400 series jobs and agencies, September 2017*

Agency	Department of the Interior						Department of Agriculture			Department of Defense		NOAA	DOE	EPA	Total
	BLM	BIA	USBR	USGS	NPS	USFWS	USFS	NRCS	APHIS	USACE	DOD Other				
0401–General Natural Resources Management and Biological Sciences	677	141	168	298	617	1,817	2,142	546	1,298	2,855	1,254	465	69	1,191	13,538
0454–Rangeland Management	291	22					304	285							902
0455–Range Technician	1,448	57			12	89	245								1,851
0460–Forestry	154	115				22	1,326	43		46	69				1,775
0462–Forestry Technician	406	427			370	156	14,440				90				15,889
0480–Fish and Wildlife Administration						251									367
0482–Fish Biology	59		31	189	37	649	281			80	6	870	27		2,243
0485–Wildlife Refuge Management						583									583
0486–Wildlife Biology	228		12	180	86	529	598	41	450	41	95	21	41		2,281
0025–Park Ranger	294	23			4,130	259	54			404	3				5,167
Total	3,557	762	234	667	5,252	4,355	19,390	915	1,748	3,426	1,529	1,433	137	1,191	44,596

Note: Explanations of agency abbreviations are given in the text.

*Three job series had much higher numbers when compared to March 2017. These were 455 Range Technician (888 in March), 462 Forestry Technician (8,910 in March), and 0025 Park Ranger (4,146 in March). The increased values for September reflect the seasonal nature of natural resources employment.

(Source: US Office of Personnel Management 2018.)

followed by the Fish and Wildlife Service (529). When it comes to range managers, the Forest Service employs slightly more (304) than the Bureau of Land Management (291).

The National Park Service is the second largest federal employer of natural resources professionals and technicians, with about 12 percent of the total. Of its 5,252 positions, over 78 percent are park rangers. The third largest employer is the US Fish and Wildlife Service, with about 10 percent of the total. About 42 percent of USFWS personnel are general natural resource scientists. The National Oceanographic and Atmospheric Administration employs the most fisheries managers (482 series). Wildlife refuge manager positions (485 series) were hired only in the US Fish and Wildlife Service.

What follows is a brief description of the agencies that employ significant numbers of natural resources professionals.

Department of Interior

Bureau of Land Management (BLM). One of the younger federal agencies, it was formed in 1946 out of the General Land Office and Grazing Service, but its origins go back to the 1780s. The Bureau of Land Management's mission is to sustain the health, diversity, and productivity of America's public lands for the multiple use and enjoyment

Public lands are open to multiple public uses, but need monitoring, planning, and management—all tasks of the natural resources professional. The Virtue Flat Off-Highway Vehicle Area in Oregon offers trails for off-road enthusiasts who purchase an all-terrain vehicle permit. (Photo by Greg Shine, BLM, Flickr)

of present and future generations. The BLM manages the most land of any US agency—about 248 million acres or 40 percent—and employs a little over 11,500 people. It has been called "the nation's largest landlord" (Skillen 2009), managing diverse land types within its units. It manages extensive grasslands and some of the Pacific Northwest's most productive forestlands, as well as small cultural sites. Its units include national monuments, wilderness areas, wild and scenic river corridors, national historic trails, and scenic trails. It manages 155 million acres of grazing land and has 63,000 oil and gas wells on its lands. With regard to the 400 job group, the BLM primarily employs range technicians, general natural resource scientists, and forestry technicians.

US Fish and Wildlife Service (USFWS). Formed in 1940 from the Bureau of Fisheries and the Bureau of Biological Survey, the USFWS claims a mission of "working with others to conserve, protect, and enhance fish, wildlife, plants, and their habitats for the continuing benefit of the American people." The USFWS is a diverse agency, managing the National Wildlife Refuge System with more than 550 wildlife refuges, and many other units with a total land area over 150 million acres. More than one-third of this area, nearly 55 million acres, is in the Pacific Ocean—administered as national monuments—and over half of the area is in Alaska. The area of the state of Alaska is nearly 18 percent wildlife refuges. The USFWS also manages the federal Endangered Species Program, Migratory Bird Management and Duck Stamp Programs, and the National Fish Hatchery System. These include 70 hatcheries, law enforcement offices, and a fish and wildlife forensics lab. Within the 400 job group, the USFWS primarily employs general natural resource scientists, fisheries managers, wildlife refuge managers, and wildlife managers.

National Park Service (NPS). The National Park Service has the challenging mission of both protecting ecological and historical resources and hosting large numbers of visitors for recreational use. The goal of the National Park Service is to preserve unimpaired the natural and cultural resources and values of the National Park System for the enjoyment, education, and inspiration of current and future generations. NPS cooperates with partners to extend the benefits of natural and cultural resource conservation and outdoor recreation throughout the United States and the world. The NPS manages the nation's 59 national parks as well as more than 75 of the nation's 129 national monuments; it shares management responsibilities for several other national monuments. The NPS has over 15 different titles for its holdings, including national lakeshores, national seashores, national battlefields, national military parks, national recreation areas, and national parkways.

There are about 400 units within the NPS, encompassing about 88 million acres. The NPS has more than 15,800 permanent employees and hires around 3,000 seasonal employees each year. The NPS employs large numbers of park rangers, as well as significant numbers of general natural resource scientists and forestry technicians.

Bureau of Reclamation (USBR). The mission of the Bureau of Reclamation is to manage, develop, and protect water and related resources in an environmentally and economically sound manner in the interest of the American public. This mission encompasses more than 150 projects and 490 dams in the western United States. It is the largest supplier of wholesale water in the United States and manages over 7 million acres of land. Within the 400 job group, the USBR primarily employs general natural resource scientists, with smaller numbers of fisheries managers, wildlife managers, and park rangers.

Bureau of Indian Affairs (BIA). The Bureau of Indian Affairs' mission is to enhance the quality of life, to promote economic opportunity, and to protect and improve the trust assets of American Indians, Indian tribes, and Alaska Natives. The BIA manages over 55 million acres in trust for Native American tribes. The BIA primarily employs forestry technicians, general natural resource scientists, and foresters.

US Geological Survey (USGS). The USGS was founded in 1879 to carry out the "classification of the public lands, and examination of the geological structure, mineral resources, and products of the national domain." It does not manage land but "serves the nation by providing reliable scientific information to describe and understand the Earth; minimize loss of life and property from natural disasters; manage water, biological, energy, and mineral resources; and enhance and protect our quality of life" (USGS n.d.). Within the 400 job group, the US Geological Survey employs general natural resource scientists, fisheries scientists, and wildlife scientists.

Department of Agriculture

Forest Service (USFS). The mission of the USFS is to sustain the health, diversity, and productivity of the nation's forests and grasslands to meet the needs of present and future generations. The Forest Service manages the second largest amount of federal land area, with 193 million acres of land in 155 national forests and 20 grasslands. It also has a forestry research program, international programs, and other functions. It is the largest employer of professionals and technicians in the 400 group; these are primarily forest technicians, general natural resource scientists, and foresters.

Natural Resource Conservation Service (NRCS). The Natural Resource Conservation Service does not manage land. Its mission is to provide resources to farmers and landowners to aid them with con-

Fish biologists conduct stream sampling for fish habitat surveys on the Siuslaw National Forest. (Photo by USFS Siuslaw National Forest, Flickr)

servation. It is most notable for its soil surveys and conservation work. Within the 400 group the NRCS primarily employs general natural resource scientists and range managers.

The Animal and Plant Health Inspection Service (APHIS). The mission of APHIS is "to protect the health and value of American agriculture and natural resources." It is involved in regulating genetically engineered organisms, administering the Animal Welfare Act, and conducting wildlife damage management. APHIS employs general natural resource scientists and wildlife biologists.

Department of Defense

The Department of Defense (DOD) includes the US Army, Air Force, Navy, and Marines. Along with the US Army Corps of Engineers (USACE, see below), the DOD manages more than 11 million acres of land on military bases and US Army Corps of Engineers projects. Separate from the US Army Corps of Engineers, the DOD civilian workforce employs about 1,200 general natural resource scientists. It also employs smaller numbers of specialists across the 400 job group.

US Army Corps of Engineers (USACE). The USACE is part of the US Army but engages in both military and civilian projects. It has about 37,000 employees and provides support to more than 150 US Army installations and over 90 Air Force installations. It operates more than 600 dams and maintains 900 harbors and 2,500 recreation sites. With regard to natural resources professionals, the USACE, separate from the DOD, primarily employs general natural resource scientists and park rangers.

Department of Commerce

National Oceanographic and Atmospheric Administration (NOAA). NOAA's mission is "to understand and predict changes in climate, weather, oceans, and coasts, to share that knowledge and information with others, and to conserve and manage coastal and marine ecosystems and resources." The NOAA does not manage land, but it does hire fisheries managers, many of them as observers on ocean vessels; and wildlife managers, particularly those who specialize in marine mammals. It also hires general natural resource scientists and fish and wildlife administrators.

Department of Energy (DOE)

The DOE does not manage wildlands, but its interest in energy generation creates jobs for wildlife and fisheries managers, as well as general natural resource scientists.

A US Army Corps of Engineers wildlife biologist monitors bird nests as part of an environmental mitigation project for dredging in the Savannah Harbor. (Photo by Tracy Robillard, USACE, Flickr)

Environmental Protection Agency (EPA)

The EPA is not a Cabinet-level department but rather is an independent agency. It does not manage land, but has a broad mission to protect human health and the environment. In pursuit of this mission, it employs a significant number of general natural resource scientists.

In addition to the above agencies, natural resource job-seekers should investigate other agencies that have specialized needs; that is, they may not hire many professionals within the traditional natural resources categories, but they may have positions that are closely related. For example, several federal agencies hire analysts and natural resource specialists with experience in geographic information systems, satellite imagery, and aerial photo interpretation.

Quasi-Government Agencies

Quasi-government agencies like the Tennessee Valley Authority (TVA) or the Bonneville Power Administration (BPA) are well known organizations. Both manage dams and hire fisheries managers. The TVA also manages more than 450,000 acres of land and hires natural resources professionals to manage its land holdings.

TRIBAL GOVERNMENTS AND ALASKA NATIVE CORPORATIONS

Tribal Governments and Alaska Native Corporations often have forests and interests that are not held in trust by the Bureau of Indian Affairs. These tribes and corporations hire foresters and other natural resources professionals to assist in managing their forests and other interests. The Intertribal Timber Council produces the *Indian Forestry & Natural Resources National Directory* (Intertribal Timber Council n.d.) that lists all tribal natural resource departments.

STATE AND LOCAL GOVERNMENTS

State Governments

All 50 states have at least one agency that enforces the state's fish, wildlife, and forest laws and this agency may also manage land areas in state parks and forests. Some states have limited land holdings while others have extensive land holdings. Be aware that states vary in how they organize and name their natural resource agencies. Some states have comprehensive departments of natural resources that are divided into separate divisions for different functions. An example is Michigan, in which the Michigan Department of Natural Resources regulates forestry, wildlife, and fisheries; manages the state parks, state forests, and game areas; regulates mining and oil; regulates boating; and has its own law enforcement branch. Other states organize their natural resource agencies quite differently. For example, Oregon has separate departments for forestry, fisheries and wildlife, parks, mining, state lands (which hold some of the state forests, but not all), a marine board that regulates boating, and a water resources department that oversees water use. In Oregon, the enforcement of fish and game laws is conducted by the state police. In Wyoming, the wildlife agency is called the Game and Fish Department; whereas in California it was called the Department of Fish and Game and changed in 2012 to the Department of Fish and Wildlife.

State natural resource agencies can be quite large. One of the largest is Cal Fire (formerly called the California Department of Forestry and Fire Protection) with over 5,000 permanent and 2,000 seasonal employees. Cal Fire operates more than 500 fire stations, 600 fire engines, and has its own fleet of 23 air tankers, 15 tactical aircraft, and 12 helicopters. The agency manages eight state forests totaling in excess of 71,000 acres. Another example of a large state agency is California State Parks, a separate agency with over 8,500 permanent staff, some 2,700 seasonal employees, and more than 275 park units. One of its parks, Anza-Borrego State Park, covers 600,000 acres. California State Parks also oversees 280 miles of the California coastline. Land areas governed by state agencies can also be large. In New York State, the Adirondack Park encompasses 9,375 square miles, with about 44 percent of that in state-owned forest preserve.

A list of the state and territorial fish and wildlife departments can be found on the websites of the US Fish and Wildlife Service and the Association of Fish and Wildlife Agencies. State forestry departments are listed on the National Association of State Foresters website. This site not only provides links to state forestry agency sites, but also includes a job site for state forestry jobs.

Local Governments

Many county and city governments also manage parks and forested areas. Some county parks, forests, and community forests are large enough to employ natural resources professionals, but many are not and rely on consultants who also work with other clients. Examples of some large forest areas under county management are the Cook County Forest Preserves that surround the city of Chicago, and the county forests of Wisconsin which total about 2.4 million acres (Wisconsin County Forest Association n.d.). Most major cities employ several urban foresters.

OTHER EMPLOYERS OF NATURAL RESOURCES PROFESSIONALS

Nongovernment Organizations

The Nature Conservancy is an example of a nongovernmental organization (NGO) that hires natural resources professionals. It is somewhat unique in that it owns more than two million acres of land and holds over three million acres in conservation leases (The Nature Conservancy n.d.).

Educational Organizations

Colleges and universities employ instructors, researchers, and technicians. Many colleges and universities have research programs and school forests for teaching and research. Some schools even have their own fish hatcheries. Land-grant universities are required to have extension services that "extend" or convey scientific research and technology to society via workshops, demonstrations, and seminars. There are forestry, wildlife, fisheries, and range extension services employing extension agents in each of these professions.

Private Landowners and Managers

It is difficult to obtain statistics on how many natural resources professionals work for private firms. In May 2017, the Bureau of Labor Statistics indicated about 1,900 foresters and 3,400 wildlife and fisheries managers and zoologists worked in private industry in the United States (Tables 4-1 and 4-2). In a salary study commissioned by the American Fisheries Society, researchers surveyed more than 900 firms that were identified as having hired fisheries employees; only 52 responded (Responsive Management 2013). These 52 firms employed 868 fisheries employees. The firms included 17 nonprofit organizations (309 fisheries employees), 26 environmental consulting firms (503 fisheries employees), 3 aquaculture/hatchery operations (27 fisheries employees), 4 power and utility companies (7 fisheries employees), and 2 firms classified as other with 7 fisheries employees.

As of 2015, there were 312 private organizations that owned or managed 10,000 acres or more of timberland; 114 of these owned or managed 100,000 acres or more (FORISK Consulting 2017). The landscape of ownership of commercial forestland has changed rapidly in the last few decades. The shift has been away from industrial ownership; that is, those involved directly in managing and processing the timber. Some of these companies were highly integrated, meaning they owned their own nurseries, timberland, harvesting crews, trucks, mills, and sales facilities. Changes in tax laws, the availability of timber on the open market, and the liability and complexity of industrial logging, trucking, and mill operations have resulted in a dramatic decrease in industrial forest owners and the shift to other forms of organization for timber ownership and management. For a better understanding of this shift, see the brief histories given in Mendell (2016) and Stein (2011).

A Real Estate Investment Trust (REIT) is a company that owns, operates, or finances income-producing real estate. REITs are becoming more common, with Weyerhaeuser now in this category of ownership. Weyerhaeuser, with over 12 million acres of timberland, is currently the largest landowner in this category and one of the largest private landowners in the nation. Another change in forestland management is the development of Timber Investment Management Organizations (TIMOs) that do not own land, but manage large land areas for investors. In 2016 the top five TIMOs managed from two to four million acres each. Forests and grasslands (ranches) can also be owned by private individuals, or families, often through privately held corporations. Many of these owners hire foresters, range managers, wildlife managements, and fisheries managers.

The changing landscape of private, commercial forestland ownership is highlighted by one forester in California who has spent his entire career at the same job, in the same office, but has worked for three different companies. Of course, not all natural resources professionals keep their jobs after ownership changes, and it is not uncommon to hear of employees with decades of experience in one company being laid off when the land they manage is acquired by another company. Versatility and the ability to build networks with other natural resources professionals in both the private and public sectors are key to maintaining an ongoing career. Note that these changing patterns of ownership affect not only foresters but also wildlife and fisheries managers. Sometimes the wildlife and fisheries managers outnumber the foresters in a private company's field offices.

SUMMARY

As of May 2017 there were about 18,000 positions classified as professional wildlife or fisheries managers, or zoologists in the United States. During the same period, professional forestry positions numbered about 8,300. In both categories the majority of jobs were in federal, state, or local governments. About 23 percent of the forestry positions and 26 percent of the wildlife, fisheries, and zoology positions were in private industry, with the remainder classified as "other." A variety of federal agencies hire natural resources professionals, including several agencies that do not manage land.

STUDENT EXERCISES

1. Choose a state in which you are interested. Determine the structure of the government agencies that regulate natural resources in the state. Determine if the agencies manage any state lands or only regulate the resources.
2. Contact the local chapter of a professional natural resource society—either the Society of American Foresters, Society for Range Management, American Fisheries Society, or The Wildlife Society, or contact the local office of a natural resource agency. Arrange to attend a meeting or find a local contact and arrange a ride-along or interview with a natural resources professional. Ask questions about his or her employer and the nature of the duties performed.

WORKS CITED

Bureau of Labor Statistics. 2018a. Occupational Employment Statistics, Occupational Employment and Wages May 2017, Foresters. Last modified March 30, 2018. https://www.bls.gov/oes/current/oes191032.htm.

Bureau of Labor Statistics. 2018b. Occupational Employment Statistics, Occupational Employment and Wages May 2017, Zoologists and Wildlife Biologists. Last modified March 30, 2018. https://www.bls.gov/oes/current/oes191023.htm.

FORISK Consulting. 2017 (May 4). *Forisk Blog, Forisk Forecast: Tracking the Top Timberland Owners and Managers in the U.S. and Canada, 2017 Update.* http://forisk.com/blog/2017/05/04/forisk-forecast-tracking-top-timberland-owners-managers-u-s-canada-2017-update/

Intertribal Timber Council. n.d. Publications and Resources. Accessed September 17, 2017. http://www.itcnet.org/resources/

Mendell, B. 2016. From cigar tax to timberland trusts, a short history of timber REITS and TIMOS. *Forest History Today* (Spring/Fall):32–36. https://foresthistory.org/wp-content/uploads/2017/10/Mendell_REITSandTIMOS.pdf

The Nature Conservancy. n.d. About Us: Private Lands Conservation. Accessed September 17, 2017. https://www.nature.org/about-us/private-lands-conservation/index.htm

Responsive Management. 2013. *2012 Fisheries Professionals Salary Survey*, Revised 4/15/2013.

Skillen, J. R. 2009. *The Nation's Largest Landlord: The Bureau of Land Management in the American West.* Lawrence: University Press of Kansas.

Stein, P. R. 2011. *Trends in forestland, ownership and conservation. Forest History Today* (Spring/Fall): 83–86. https://foresthistory.org/wp-content/uploads/2016/12/2011_Trends_in_Forestland_Ownership.pdf

US Geological Survey. n.d. About us. Accessed April 30, 2018. http://www.usgs.gov/about/about-us

US Office of Personnel Management. 2018. FedScope, Employment Cubes. Accessed April 2018. https://www.fedscope.opm.gov/ibmcognos/cgi-bin/cognosisapi.dll

Wisconsin County Forest Association. n.d. Home page. Accessed September 17, 2017. http://www.wisconsincountyforests.com/

Education

Chapter Outline
Academic Definitions
Natural Resources Majors
Components of an Academic Curriculum for a Bachelor's
    Degree in Natural Resources
Accreditation
Professional Certification
The Future of Natural Resources Education
Continuing Education
Top Ten Tips for Potential Students in the Natural Resources
Summary

The best college program for a student is the one that best fits his or her individual needs! It should be a good fit in terms of the academic degree and emphasis, the cost, and the location. If a student knows exactly what profession he or she wishes to pursue, then a program that leads directly into that profession is a great choice. What is more challenging is when students do not know which natural resource profession they wish to pursue. In such cases, a college that offers several natural resource programs may be the best option. Some students are place bound, with their options constrained to local schools. Others are free to travel, but are academically or financially constrained. Some students thrive in a high-pressure, formal environment at a large, prestigious, research-oriented university; others do better in a small, laid-back school. Each program in each school will have its own culture; similarly, each cohort of students will have its own characteristics. Note that even within large universities, the traditional natural resources programs may have modest enrollments, giving faculty and students an opportunity to become familiar with one another. In any traditional natural resource program, students will likely have classes and labs with most of the instructors in the program, and will work closely with one or two of those instructors during their time as a student.

This chapter starts with a few academic definitions and then progresses into the specifics of natural resource programs in forestry, wildlife management, fisheries management, and range management.

Academic Definitions

The following broad definitions will help readers distinguish among the different types of degrees offered by various institutions for higher education.

Major. A specific subject in which a student specializes.

Certificate. Certificates are earned for completing a required set of courses.

Associate's Degree. An associate's degree is designed to be completed in two years for a well-prepared student. Many students take longer than two years. There are many types of associate degree, but the most common are the Associate of Applied Science, the Associate of Science, and the Associate of Arts.

An Associate of Applied Science (AAS) degree is designed to prepare a student for a career as a technician within a field such as forestry. It will require coursework mainly in the subject area but also in communication and mathematics. An AAS degree from a community college is not designed to be a transfer degree, and pursuing an AAS with the goal of completing a bachelor's degree will be inefficient. However, if employers are familiar with the requirements of an AAS degree, and all other factors being equal, they may preferentially hire a candidate with both the practically oriented AAS and a bachelor's degree.

The Associate of Science (AS) degree is designed to prepare a student to transfer to a specific college's major program of study to obtain a bachelor's degree. This is a good option for those students who have determined to which college they will transfer and the major they will pursue. The AS degree will have coursework in the subject area, but a student will also take courses to acquire a breadth of knowledge in multiple subject areas in what is often referred to as the baccalaureate core.

An Associate of Arts (AA) degree is designed to allow students to explore learning and take a variety of courses while completing basic courses in communication, math, sciences, and humanities. This is a good degree for students who have not yet decided where they want to study or which major they will pursue. But for those students who already know which college or university they will transfer to as well as the field in which they will major, the Associate of Arts degree often requires more credits in the humanities and social sciences than is necessary to obtain a bachelor's degree, and thus involves additional time and expense to complete a college degree.

An associate's degree generally requires at least 60 semester hours, or 90 quarter hours, of credit to complete the degree. An associate's degree in a natural resources discipline often requires *more than* 60 semester hours, or 90 quarter hours, to assure adequate preparation for a career. Most states set an upper limit around 70 semester hours or 105 quarter hours for an associate's degree, and the credits required by many associate's degrees in natural resources will be at, or near, their state's upper limit.

Bachelor's Degree. Bachelor's degrees are designed to be obtained in four years by a diligent and well-prepared student. The bachelor's degree requires a student to obtain both a breadth and a depth of knowledge in their major subject. The most common degrees are the Bachelor of Science (BS) and Bachelor of Arts (BA). As the names imply, the focus of the degree differs, although this varies by college. Some colleges grant a Bachelor of Arts to science majors, particularly colleges that require both a wide range of core courses and depth of knowledge in the specific science topics. Most bachelor's degrees in natural resource subjects will be Bachelor of Science.

A bachelor's degree generally requires at least 120 semester hours, or 180 quarter hours, of credit. A bachelor's degree in natural resources often requires *more* credits to assure adequate preparation for a career. The upper limit for credits is around 150 semester hours or 225 quarter hours as determined by the state or college's rules. It is common to find natural resources major programs of study hovering near the highest credit loads at many universities.

Master's Degree. Master's degrees are designed to be two, or more, years in length and allow a student to complete an in-depth study of a specific subject. Master's degree programs generally require the completion of a bachelor's degree before admission to the master's program. Most master's degrees require the completion of a project or a thesis. A master's thesis is an in-depth, written discussion completed after researching a relevant topic. A master's project requires an in-depth report on some type of observational or experiential activity. There are many types of master's degrees; the most common are the Master of Science (MS) and the Master of Arts (MA). The most common degree in the natural resources disciplines is the Master of Science, although there are specific degrees such as the Master of Forestry (MF). In natural resources disciplines, completion of a master's degree often takes three or more years.

Doctorate Degree. The most common research-based doctoral degree is the Doctor of Philosophy (PhD), which requires independent in-depth research and a dissertation, which is an extensive, written discussion of the research. This is the most common doctorate held by natural resources professionals. Commonly, it takes five to seven years or more to complete a doctorate. Requirements vary by school, but within the natural resources disciplines a master's degree is usually required for admission to a doctoral program.

Professional Degree. These typically refer to degrees in the fields of medicine, law, and divinity (ministry). There are few professional degrees in natural resources.

Terms. Colleges in the United States generally use one of three systems for setting their academic calendar and designating the term or length of a course. Some use the quarter system of four terms (fall, winter, spring, and summer) of about 11 to 12 weeks each, with the summer term being optional and often shorter. Many use the semester system of two primary terms (fall and spring) of about 16 to 18 weeks each, and a shorter, optional summer term. Some institutions add short (three or four week) terms between the fall and spring term or between the spring and summer term. These short terms allow for intense, focused courses. In the third system, the academic year is divided into trimesters of 14 to 16 weeks each, with students normally attending two of the three terms.

The typical natural resource courses are not usually offered during the summer term due to the expectation that students will be employed in field jobs or attending special field courses or summer camps. For natural resource employers and students, the quarter system meshes well with the field season and wildfire season in the western US. The fall term of a quarter system starts around the end of September and the spring term ends in mid-June; a near match with the fire season in the western US. The summer travel and recreation season meshes well with the semester system in which the spring semester ends in mid-May, allowing students to begin work and training in parks and recreational areas before the summer recreational season peaks.

Credits. Colleges vary somewhat on how they award credit, but generally, a course is assigned one credit for each hour of traditional lecture per week, and one credit for every three hours of a traditional lab per week. Tutorials, discussions, and combination lecture/labs will be apportioned credit somewhere between the credits given for traditional lectures and labs. Credit hours awarded under a quarter system are worth 2/3 that of a semester credit hour.

Curriculum. The topics and subjects taught in a course or a major program of study.

Community or Junior College. An institution offering associate's degrees, certificates, and occupational training. Most often these are public institutions within a district and governed by a locally elected board and state rules. A few community colleges are private institutions.

College. An institution offering, at least, bachelor's degrees. There are both private and public colleges in the United States, although those offering forestry, wildlife management, fisheries management, and range management are generally public institutions and usually limited to a few within a region.

University. An institution offering, at least, master's degrees. Most universities are organized into separate schools, or colleges, that specialize in a subject area—such as a university with a college of science and a college of arts. Many natural resource degree programs in the United States are housed in state, land-grant universities that were financed by selling land granted to the state by the federal government. In exchange, the land-grant university is required to teach military science, agriculture, and other technical subjects. Further, such universities are required to conduct agricultural research and disseminate the results in a practical manner. These requirements are the origin of the Agricultural Extension Services 4H Programs and have resulted in forestry and agricultural extension services throughout the western United States.

College or University Catalog. A college catalog is more than a list of courses and programs; it is the basis of a contract between a student and the institution. The catalog lists the requirements for a program of study, and both the institution and the student have obligations to those requirements. Since requirements may change as a student progresses through an institution, students are often given some choice of catalog requirements under which they can graduate. Institutions generally allow students to use the current catalog requirements or those from their entry year if the student is within a reasonable period from entry (five or six years). Generally, institutions will reset the catalog requirements to the time of re-entry for students who drop out or take long leaves of approved absence from the institution.

NATURAL RESOURCES MAJORS

The four traditional majors in natural resource programs are forestry, wildlife management, fisheries management, and range management. Within these categories, there may be majors that focus on a specific aspect of the field—for example, urban forestry. Newer majors include natural resources, environmental studies, and environmental science.

Within most traditional programs in forestry, wildlife management, fisheries management, and range management, students can expect a curriculum that allows them to meet the federal educational requirements for those job series (see Chapter 2). Programs that are more general, for example a broad natural resource program, should enable a graduating student to meet the 401 series requirements, and may allow the student to meet the requirements of other jobs within the 400 series (see Chapter 2 for the federal 400 series requirements for forestry, wildlife management, fisheries management, and range management positions).

Both the traditional and newer majors may be called something different at each college. In some cases, a major may be given a certain name to reflect that college's particular emphasis; in other instances, majors may be given catchy names to attract students. Any institution of higher learning in the United States that accepts federal funds must classify its majors with a Classification of Instructional Programs (CIP) code; a code used in the federal education system database. There are 24 CIP codes used to classify college natural resource programs in the United States (Table 5-1). The college's title for the major or the CIP code may not describe the exact content of the major, but the CIP will be used by websites for searching and sorting programs by type of major. The CIP also provides a standardized code to compare the number of institutions offering that major.

COMPONENTS OF AN ACADEMIC CURRICULUM FOR A BACHELOR'S DEGREE IN NATURAL RESOURCES

Forestry, wildlife management, fisheries management, and range management are all science-based disciplines, heavily dependent on the science of ecology. Other sciences like biology, mathematics, statistics, chemistry, climatology, geography, economics, political science, and sociology form the foundation for ecology or influence it. Nonscientific knowledge is also important; law is one example of a nonscience topic that influences the management of natural resources. But, by far, the most important influence on the natural resource disciplines is ecology. Foresters, wildlife biologists, fish biologists, and range managers can all be thought of as applied ecologists. However, as noted in earlier chapters, the management of forests, wildlife, fish, and grasslands is carried out within the framework of a society and its values and laws. Thus, the social sciences play a role in preparing students for employment in these professions.

Education

Table 5-1 Classification of Instructional Programs (CIP) Codes for Natural Resources and Conservation Programs

03)	NATURAL RESOURCES AND CONSERVATION	Number of Institutions
03.01)	Natural Resources Conservation and Research	500+
03.0101)	Natural Resources/Conservation, General	127
03.0103)	Environmental Studies	500+
03.0104)	Environmental Science	500+
03.0199)	Natural Resources Conservation and Research, Other	27
03.02)	Natural Resources Management and Policy	160
03.0201)	Natural Resources Management and Policy	93
03.0204)	Natural Resource Economics	15
03.0205)	Water, Wetlands, and Marine Resources Management	26
03.0206)	Land Use Planning and Management/Development	142
03.0207)	Natural Resource Recreation and Tourism	8
03.0208)	Natural Resources Law Enforcement and Protective Services	9
03.0299)	Natural Resources Management and Policy, Other	93
03.0301)	Fishing and Fisheries Sciences and Management	30
03.05)	Forestry	131
03.0501)	Forestry, General	62
03.0502)	Forest Sciences and Biology	21
03.0506)	Forest Management/Forest Resources Management	25
03.0508)	Urban Forestry	15
03.0509)	Wood Science and Wood Products/Pulp and Paper Technology	14
03.0510)	Forest Resources Production and Management	5
03.0511)	Forest Technology/Technician	35
03.0599)	Forestry, Other	15
03.0601)	Wildlife, Fish and Wildlands Science and Management	98
03.9999)	Natural Resources and Conservation, Other	32

Note: Despite what the institution names the major program, if it accepts federal funds, the program will be classified under one of these codes.

Source: National Center for Education Statistics n.d.

One job of natural resources professionals is to research the condition of the resource. Here, a Bureau of Land Management employee gathers vegetation data using a sampling frame to define the area of the sample. (Photo by BLM, Flickr)

It is important to distinguish between different types of science. One classification is natural sciences versus social sciences. Natural sciences are those that investigate the natural world, such as physics, chemistry, and biology. Social sciences investigate society and the relationships among people, such as sociology, archeology, and economics. Another way to classify science is as pure, empirical, applied, or interdisciplinary. Mathematics and logic are *pure sciences* in that they can be entirely conceptual, without regard to practical thought. Physics is an *empirical science* because it can be observed; chemistry and biology also are empirical sciences. Physics is the foundation for the *applied science* of engineering. Forestry, wildlife, and fisheries are *interdisciplinary sciences* dependent on the pure science of mathematics, but also on physics, chemistry, and biology. Interdisciplinary sciences are removed from the pure sciences and will be the least exact and most open to interpretation. Indeed, the management of natural resources is not exact, but rather is imprecise and complicated and relies on both the natural sciences and the social sciences. Recall the statement in this book's introduction: natural resource management is not rocket science, it is much more complex!

The role of professionals who manage natural resources is to research the condition of the resource, interpret the "purer" sciences, and then apply that interpretation to take action to change or maintain the resource within the context of societal values. This can mean many compromises and trade-offs between different values and concerns. As an example, it can be very frustrating for a forester to allow a million dollars of timber to burn in a wildland fire because resources were shifted to protect a house worth half a million dollars. Simple economics would favor saving the timber; so too would the ethic of preserving wildland resources over the built environment. However, our society values personal property and we empathize with homeowners and place pressure on managers to prevent that loss. Similarly, a wildlife biologist may encounter resistance to placing an area off limits to motorized traffic to protect an animal's breeding ground because the public may place a higher value on maintaining motorized access than on protecting the breeding ground. Natural resource managers operate within this social sphere and need to be cognizant of the social ramifications of their decisions. This is particularly true as professionals advance in their careers and begin to influence policy decisions. Hence the education and training of natural resource managers must go beyond the pure and empirical sciences.

Each college or university has its "core" courses—the institution's foundational requirements for all students that assure basic competency and breadth of learning. Usually the core will include communication, both written and verbal; math; humanities; foreign language; social sciences; and natural sciences. Other common core requirements

Academic preparation for a natural resource career requires outdoor fieldwork. The late Professor Burt Barnes of the University of Michigan was a master at teaching field labs. (University of Michigan SNRE, Tribute to Burt Barnes, Flickr)

include diversity studies, writing in the context of an advanced subject, and computer literacy.

In addition to the core courses, many majors require students to take another set of foundational courses that apply specifically to the major program of study. For example, a college that offers range, forestry, and wildlife programs may have a common set of fundamental courses, or "natural resource core" courses applicable to all three majors. Basic statistics for natural resources is an example of a course that could serve all three majors. Other foundational courses may be shared with science programs, such as a basic ecology course. Some foundational, and sometimes advanced, courses may simultaneously meet the requirements for both the college core and the program requirements. Careful planning and consultation with your advisor will prevent duplication of core requirements. For example, a student may take a sociology course to meet the university's core requirement only to later discover he needs to take a forest sociology course to meet the forestry program requirement, and the forest sociology course would have also met the core requirement for sociology. Although no educational experience is a complete waste of effort, meeting a requirement twice due to poor planning is an inefficient use of time and resources.

Students will also be required to take advanced courses in their major program of study. Students may not be allowed to take these advanced courses until they have completed the fundamental courses; sometimes there is a formal admission or matriculation process to enter the advanced portion of the program of study. In advanced courses the work will become more focused, and a cohort of students progressing through the major at the same time will likely develop. This may be a formal cohort of students officially admitted to the major and assigned a cohort, or an informal group that happens to be at the same stage of the major. Students will likely become familiar with most of the students in their cohort when taking the advanced courses in their majors.

Finally, there will likely be a final "capstone" course that draws on all the knowledge and skills taught in the program. Such capstone courses often culminate in an individual or group project.

ACCREDITATION

Most colleges and universities are accredited through a regional accreditation authority that accredits the entire school. Some natural resource programs are additionally accredited; for example, the Society for Range Management (SRM) accredits degree programs in range management, and the Society of American Foresters (SAF) accredits associate, bachelor, and master degree programs in forestry, urban forestry, and natural resources and ecosystem management. The accreditation standards from the Society for Range Management and the Society of American Foresters are provided in Appendix A. The advantage of enrolling in an accredited program is that students are assured that the program meets standards for faculty qualifications, class size, safety, and curriculum content. Accredited programs must assure their accreditors that they (1) have clearly stated outcomes, (2) assess and evaluate their courses, (3) have ongoing improvement plans for programs and faculty, (4) have ample financial and administrative support to remain viable, and (5) produce graduates who are well prepared for the profession and meet standards for registration, licensing, and certification. A discussion of the benefits of accreditation is provided in Redelsheimer et al. (2015). The disadvantage of accreditation is that the program is more rigid, with less flexibility for individual students.

PROFESSIONAL CERTIFICATION

Many natural resource programs, such as wildlife and fisheries, are not accredited beyond the institution's regional accreditation. Instead, the professional society that unofficially governs the profession offers certification to individuals who have taken the appropriate college courses and have adequate work experience within the profession. The Wildlife Society, American Fisheries Society, the Society for Range Management, and the Society of American Foresters all offer certification to individuals who meet their society's criteria. Standards for certification by these societies are provided in Appendix B. The focus of certification is on individuals who meet the society's criteria, not the college program in which they participated. The advantage to this method of credentialing is that individuals can obtain their natural resource education in a variety of ways and at a wider variety of schools. The disadvantage is that an individual may not be assured that a specific program has courses that meet the criteria for certification. Certification is flexible and versatile for an individual, but somewhat more "buyer beware" than accreditation. For example, a college may have a natural resources program with a fish and wildlife emphasis. It may be an academically rigorous program, but the degree holder may not be able to meet the requirements for a federal job in fisheries or wildlife (see Chapter 2) without taking additional courses in either fisheries or wildlife. Students and their advisors bear more responsibility for the student's educational program with individual certification compared to accredited programs.

THE FUTURE OF NATURAL RESOURCES EDUCATION

Students in a traditional natural resource program should realize that these programs are not static, and there are ongoing discussions about how to improve them, the appropriate role of distance education, and the content of their curricula. There have been several symposiums on natural resource education, including one that was sponsored by the American Fisheries Society as part of its 2009 annual meeting entitled *Fisheries Education in the 21st Century: Accommodating Change*; the Coalition of Natural Resource Societies Natural Resource Education and Employment Conference in 2011, which produced a report published in 2012; and the University of California—Berkeley's North American Summit on Forest Science Education in 2014.

One concern is producing well-trained individuals who are both ready for the workplace and "society ready." Society ready is a term coined by Bullard et al. (2014) to describe a forester who can deal effectively with the social, economic, and ecological issues that he or she will confront. The term could apply to any natural resources professional. The premise of the article was how to use evidence to change a college curriculum to improve personal skills and people skills, what some call "soft skills" and others call "human dimensions." The article's authors describe how they changed their college program's curriculum from one that emphasized technical skills first, then general

Education

skills, and finally personal skills to a program that placed equal emphasis on all three skill sets.

Another area of concern is declining enrollments. Sharik et al. (2015) analyzed enrollments in natural resource programs in institutions with membership in the National Association of University Forest Resources Programs (NAUFRP). They found that enrollment patterns in natural resource programs from 1980 to 2009 were similar in all parts of the country, indicating that national rather than regional issues influence enrollments. Enrollments declined by about 50 percent from the early 1980s to 1990. They then began to increase until peaking in the mid-1990s at levels similar to those in 1980. This was followed by a decline, but enrollments have been slowly increasing since 2005–2006.

Traditional forestry programs have been the most volatile. From 1980 to 2009, their enrollment declined from 47.3 percent of total natural resources enrollment to 22.0 percent. During this same period enrollments in fish and wildlife programs increased from 15.6 percent of the total to 23.1 percent, and general natural resources and environment program enrollment increased from 14.5 percent to 35 percent. The other programs' enrollments remained relatively stable. The proportion of female students in natural resource programs increased during this same period from 34.5 percent in 1980 to 40.8 percent in 2009.

Sharik and coauthors credit several trends that may have impacted these enrollments. One influence may be a shift in public attitudes

Field data may still be recorded in written form. Neat, readable printing is still a useful skill. (USFS, Flickr)

toward forestry, particularly the perception of forestry as primarily consumptive and students wishing to avoid careers in professions perceived to be consumptive. This is consistent with the increasing enrollments in fish and wildlife programs, which train students for professions that are perceived to be less consumptive than forestry. Another reason cited for the change in enrollments is the diversification of natural resource programs beyond that of traditional forestry. Historically, all of the wildland management disciplines were taught as forestry (see Chapter 3), but as wildlife, fisheries, and range management began to mature as professions, colleges developed programs for these specific professions and the "market share" of enrollment for forestry programs declined. Additionally, the authors cite the perception by many students that the general natural resources and environmental programs are less academically demanding and require fewer science and math courses. A survey of potential students indicated that their hesitancy to enter forestry programs was due to a dislike of highly quantitative subjects. Additionally, students are aware that accredited forestry and range management programs have more rigid curricular requirements. Students also perceive a lack of forestry jobs and low wages. Finally, there is a limited attraction to forestry by women and minorities.

The article by Sharik et al. was published in a special edition of the *Journal of Forestry* in November 2015. The issue reported on the University of California—Berkeley's North American Summit on Forest Science Education held in May 2014. The topics discussed at the summit included the extent to which graduates were prepared for the workforce, enrollment trends, distance learning opportunities, curriculum, master's degree options for foresters, accreditation, employment trends, international education, diversity, forestry education at research universities, and students' perspectives on their forestry education.

Similar concerns are mirrored in a report from the Coalition of Natural Resource Societies (2012) on the four traditional natural resource disciplines. The report cited evidence that traditional wildlife programs affiliated with the National Association of University Fisheries and Wildlife Programs (NAUFWP) are now a minority of all the wildlife programs offered by colleges in the United States. The 2012 report also noted that college programs in the natural resources were offering less field experience, resulting in students transferring out of the programs due to disappointment at not having extensive outdoor labs. The coalition report also cited student anxiety over the availability of jobs when they graduate; however, it also noted that natural resource agencies were claiming to have too few qualified candidates for their job openings. In particular, applicants from general natural resource programs lacked training in technical skills, while recent hires lacked communications skills and overrated their problem-solving and decision-making abilities.

The Future of the Wildlife Profession, a 2009 report from a committee of The Wildlife Society, examined 3,413 collegiate natural resource programs in the United States that offered courses related to wildlife (Baydack 2009). Few programs required students to meet The Wildlife Society's certification requirements; others offered the appropriate courses for certification but did not require them for graduation. Only about 14 percent of the programs offered wildlife curricula with traditional programs and membership in the National Association of University Fish and Wildlife Programs. One observation of the study was that students have changed from those with "direct experience via muddy boots and chore-calloused hands" to those who "have acquired more pseudo-experience of nature via television programs and other media." Challenges to wildlife programs included changes in curricula driven by declining budgets, increased cost of field labs, increased general education requirements, and tension between "theory vs. practice" and "coursework vs. experience-based learning."

The Wildlife Society report also summarized a survey of employers that identified both oral and written communication and teamwork skills as being deficient in wildlife graduates. Surveyed employers desired a wide set of skills, but perceived that recent graduates lacked basic experience and practical field knowledge. Their concept of the "ideal" wildlife program was: core competencies in biological, physical, and quantitative sciences; humanities; communication; and policy, administration, and law. Other key areas for emphasis included: teamwork and working with stakeholders, field experience, and critical thinking. Building out from this core would incorporate competency in geospatial analysis, quantitative science, behavior, ecology and evolutionary biology, and conservation biology. The authors of *The Future of the Wildlife Profession* acknowledged that this ideal would be difficult to obtain with diminishing resources, increasing college core requirements, research requirements, and "too much to know, too little time" for students in the programs. The authors concluded that both employers and wildlife professionals should be aware that continuing education for wildlife managers will become increasingly important.

The American Fisheries Society sponsored a symposium as part of its 2009 annual meeting entitled *Fisheries Education in the 21st Century: Accommodating Change*. The symposium addressed the need to change educational preparation in the context of a dynamic profession that has evolved over the last century. It acknowledged increased demands on the curriculum of fisheries programs while such programs also confronted reductions in faculty, increasing workloads, and funding cuts. The published report of the symposium acknowledged that the mission of academic programs had shifted from traditional fisheries to broader coursework to accommodate changes in employer and student expectations. It discussed how to improve students' critical and creative thinking skills, communication skills, the retention of science-based information, the diversity of students, and the generational differences between faculty and entering students.

The above discussion indicates that the traditional natural resource societies and the instructors within natural resource programs are concerned about the quality of the programs, and that these programs are evolving in response to employer and student expectations. There are continuing efforts to improve natural resource programs and their curriculums to address the needs of the professions and the need of employers to have "society ready" graduates who are properly equipped for the workplace with both knowledge and field skills.

Continuing Education

Participation in continuing education is an important characteristic of a professional. Continuing education is required by most employers and many have minimum requirements for the number of hours employees must complete. The professional societies have strict guidelines for the types and amount of continuing education that is required to renew a certification (see Appendix B). Quoting from a recent report on a congressional bill, "Professional societies, such as the Society of American Foresters, the Society for Range Management, The Wildlife Society, American Fisheries Society, and others provide opportunities for employees to maintain professional competencies through continuing education, scientific journals, and interaction with other professionals" (House of Representatives 2017). Even established natural resources professionals will be expected to participate in continuing education until the end of their careers.

Top Ten Tips for Potential Students in the Natural Resources

1. Assess whether you are prepared for college and consider delaying your start until you are fully ready.

   Academically ready. Are you prepared with foundational courses? For natural resources this would include math, biology, chemistry, and other physical sciences as well as communications. One of the best predictors of college success is the ability to read with comprehension of the material.

Financially ready. Can you afford tuition, books, materials, and room and board; but also transportation, equipment (boots and field clothes), and social expenses? What, if any, financial aid is available?

Practically ready. Do you have adequate transportation, time, and freedom from family and job obligations to commit to attending class and studying?

Behaviorally ready. If you absorbed the information in Chapter 2 on discipline, you realize your career is going to begin with your education. Be aware that the professional natural resource community is rather small. Your abilities and behavior will be noticed.

2. Do not choose a school primarily for the recreational activities offered in the area. Very few students who have "majored" in hunting, recreational skiing, or canoeing thrive academically. This does not mean students should not have fun and explore their surroundings—which is important—but academic achievement should be their top priority.

   Students commonly choose a school based on their perception of the experience they will have at that school. This can be observed in college recruiting advertisements, which are selling a college experience, but not necessarily an academic one. It is typical to see ads for colleges with pictures of smiling students engaged in some fun activity. Remember that colleges need to fill their classrooms to stay in business and they use sophisticated marketing to sell themselves to prospective students. You need to look beyond the pretty marketing photos and evaluate the school's natural resources program. Ask recruiters how the program is of benefit to students preparing for the workplace. Ask about its accreditations, advising methods, general culture, and how students interface with faculty and advisors. Recent alumni of a program are often the best source of information about the program and how well it prepared them for a career in natural resources.

3. Don't procrastinate; application deadlines for admission and scholarships are often sooner that many students and their families anticipate.

4. If you know your career path, great; focus on that path while attempting to become as versatile as possible within the framework of your selected program. However, consider giving yourself some leeway to alter your path; the career you think you want may not be a viable choice when you finish school. Agencies, companies, and personal and family situations may change, and unexpected opportunities may arise. Seek flexibility to accept

both challenges and opportunities. If you have not selected a career path, start thinking about a path and work with your advisor to efficiently explore your interests.
5. Find an advisor you can work with and heed his or her advice. If you are allowed to choose your own advisor, do so wisely; seek out someone who asks you difficult questions about your career plans, academic goals, and how well you are tracking toward those goals. Avoid frequently switching advisors to manipulate the advising process.
6. Prerequisites and course sequences exist for a reason. Pay attention to them. If you think you may have the experience and academic maturity to succeed in a course without the prerequisites, politely ask your advisor and the course instructor to enter the course. Do not attempt to simply "grab" courses that build a schedule that is conducive to recreation or your work schedule. Take courses in a progression that builds upon foundational courses. If you read through the accreditation standards in Appendix A, you will notice that the standards require accredited programs to require students to progress from foundational courses to advanced courses. Students often attempt to put off unpopular course such as math and writing until the third or fourth year of a bachelor's program. That is too late! Take them early so you can use what you learned in the advanced courses. Your advisor is remiss if he or she doesn't tell you this; see tip five above.
7. Network with other students through clubs and student chapters of professional societies. Ask their advice about instructors, courses, and programs. Some instructors have idiosyncrasies and habits that you will learn about from other students. Don't assume that other students are always correct, however; sometimes you may like an instructor that they don't, but it is good to have their perspective.

    Natural resource programs generally have small enrollments and extensive field labs—this means you will spend a great deal of time with your instructors and fellow students. You will likely form long-lasting relationships with some of these faculty members and fellow students. Graduates of natural resource programs are also amazingly loyal to their schools. Observe a gathering of natural resources professionals and the conversation will eventually pivot to which schools they attended and when; for better or worse, this information will establish their pedigree. Your career may entail working with, or for, fellow students, or they may work for you. Your reputation and relationships at school will follow you in a natural resource career.

8. Similar to tip seven above, join a professional society. Most of the professional societies have student membership rates so the financial cost is not very high. Consider the cost an investment in your career. Attendance at local, state, and national meetings will be a learning experience and yield insight into the profession and provide opportunities for professional growth. The American Fisheries Society even encourages high school students to join its society.
9. Treat your college experience like you would a job. Be on time and prepared for classes, allot enough time for study, and act responsibly and respectfully. Again, your reputation in the profession will begin in college.
10. Transferring between institutions may be beneficial and is becoming very common, with a third of students now transferring between schools (*Chronicle of Higher Education* 2012). However, such a transfer is seldom seamless and without difficulties, especially without adequate planning. But if a transfer appears to be beneficial for your particular situation, careful planning and consultation with your advisor may result in a near seamless process and full transfer of credits.

Summary

There are many paths through the academic preparation necessary for a career in natural resources. Students may start fulfilling their natural resource educational requirements at a community college or a university undergraduate program. Guidance through the comprehensive education needed for a career in the traditional natural resource disciplines may be gained through accredited educational programs or professional certification programs. Ideally, an education in natural resources both prepares a person for life-long learning and launches him or her into a career. This chapter ends with the same statement it started with; the best college program for a student is the one that best fits his or her individual needs!

STUDENT EXERCISES

1. Use the US Department of Education website (https://nces.ed.gov/collegenavigator/) to find the colleges in your area that offer forestry, wildlife, fisheries, and range management programs. The search will key on the CIP code majors given in Table 5-1. List the programs by major and distance from your location.
2. Choose three colleges from the list you created above and use the College Board website (https://www.collegeboard.org) to compare the costs of attending colleges in your region. Investigate any state-to-state exchange plans that would reduce your tuition at an out-of-state school.
3. Select a school with a natural resource program you would consider entering. Use the school's website to investigate the program's requirements. Write an academic plan for the courses you would need to complete the program.
4. Select a natural resource profession and write a description of the specialties within that profession.
5. Determine if there is a local chapter of the professional society related to your career interests. If so, attend one of its meetings.

WORKS CITED

American Fisheries Society. 2009. *Fisheries Education in the 21st Century: Accommodating Change.* Symposium presented by the Education Section of AFS at its Annual Meeting. Accessed September 21, 2017. https://education.fisheries.org/education-links/fisheries-education-in-the-21st-century-accommodating-change/

Baydack, R. 2009. The Wildlife Society Ad Hoc Committee on Collegiate Wildlife Programs. Accessed September 15, 2017. http://www.wildlifeprofessional.org/Documents/Ad_Hoc_Collegiate_Wildlife.pdf

Bullard, S. H., Williams, P. S., Coble, T., Coble, D. W., Darville, R., and Rogers, L. 2014. Producing "Society-Ready" Foresters: A Research-Based Process to Revise the Bachelor of Science in Forestry Curriculum at Stephen F. Austin State University. *J. For.* 112(4):354–360.

Chronicle of Higher Education. 2012. A Third of Students Transfer Before Graduating, and Many Head Toward Community Colleges. Accessed September 15, 2017. https://www.chronicle.com/article/A-Third-of-Students-Transfer/130954

Coalition of Natural Resource Societies. 2012. Natural Resource Education and Employment Conference Report and Recommendations. *Fisheries* 37(6):277–284.

House of Representatives. 2017. 115th Congress, Report 115-238. *Department of The Interior, Environment, and Related Agencies Appropriations Bill, 2018.* Accessed September 15, 2017. https://www.congress.gov/115/crpt/hrpt238/CRPT-115hrpt238.pdf

National Center for Education Statistics. n.d. CIP 2010 Natural Resources/Conservation, General. Accessed September 15, 2017. https://nces.ed.gov/ipeds/cipcode/cipdetail.aspx?y=55&cip=03.0101

Redelsheimer, C. L., Boldenow, R., and Marshall, P. 2015. Adding value to the profession: The role of accreditation. *J. For.* 113(6):566–570.

Sharik, T. L., Lilieholm, R. J., Lindquist, W., and Richardson, W. W. 2015. Undergraduate enrollment in natural resource programs in the United States: Trends, drivers, and implications for the future of natural resource professions. *J. For.* 113(6):538–555.

ADDITIONAL RESOURCES

Forest Science Education Special Edition. *J. For.* 113(6). November 2015. Society of American Foresters, Bethesda, MD.

2017 Guide to Forestry and Natural Resources Programs. Society of American Foresters, Bethesda, MD.

6

Practical Matters

Chapter Outline
Preparing for a Career
    Gaining Experience
    Applying for Jobs
    Interviewing
    Providing References
    Tips from a Career Guide
Human Dimension Skills
    Communication
    Ethics
    Basic and Professional Etiquette
    Working in Groups
    Records and Accounting
Natural Resources Field Skills
    Local Ecosystem Characteristics
    Basic Land and Water Navigation
    Outdoor Safety and Efficiency
Summary

Formal education is extremely important in developing a career in a natural resource profession, but field experience, practical skills, and practical knowledge are also very significant. In the previous chapter's discussion on the future of natural resource education, one of the concerns expressed was that incoming students lacked field skills. The discussion later in the chapter will summarize some of the field skills that aspiring natural resources workers should obtain. This summary is by no means exhaustive and it is not comprised entirely of field skills per se, but includes other practical skills as well. One common characteristic of natural resources professionals is that they are practical people who have learned to function effectively in a variety of environments. Not all natural resources professionals work outdoors, but those who do work outdoors efficiently have mastered the skills they need to carry out the requirements of their jobs, even in difficult conditions.

Preparing for a Career

Gaining Experience

Practical experience is an important component of job readiness. Employers desire employees who have the necessary skills to complete job assignments. Moreover, they want to hire people with the ability to function appropriately in the workplace environment, whether it is an office or in the field. Finally, by pursuing as much relevant experience as possible, prospective employees are in a better position to determine if they enjoy the work enough to make it a career.

If you are under 18 years old, seek internships or volunteer experiences. One example is the Hutton Junior Fisheries Biology Program (n.d.), which is open to high school juniors and seniors. Selected high school students are placed with mentors who are fisheries professionals. It is a prestigious program with selective admission, but worth applying to if you are interested in fisheries. Another example is the federal Youth Conservation Corps, which is a summer youth employment program that engages 15–18-year-old youth in meaningful work experiences on national parks, forests, wildlife refuges, and fish hatcheries.

Once you are 18 or older, attempt to obtain paid work experience. Unless an internship offers the exact experiences or skills that you wish to obtain, paying jobs are generally considered better than internships within the traditional natural resources disciplines. If you do have a job, consider volunteering for conservation organizations to supplement your experience.

Applying for Jobs

There are many good guides to applying for a job. One excellent resource is *A How-To Guide for Pursuing a Career in Natural Resources* (Colorado Alliance for Environmental Education 2018). Even though the guide is written specifically for jobs within the state of Colorado, it nevertheless offers excellent advice on applying for a natural resource position anywhere in the country. It also features an entire chapter on USAJOBS, the official federal job website. The USAJOBS website provides instructions for using the site to apply for a job with the federal government. Another good source of information is people who have used USAJOBS to find employment. Ideally, try to talk to someone who has experience in using the site to apply for a job, as well as someone who participates in the hiring for an agency. This is likely someone who has a few years of experience in a supervisory position. Such a person may have invaluable advice on both aspects of the hiring process. If you are a member of a student group or a professional society, invite this person to speak to your group of potential USAJOBS users.

The same advice holds true for applying to state and local government agencies or private companies. Try to find someone who knows the application process as both an applicant and as the person doing the hiring. Don't ask for inappropriate or "insider" information, but be frank and ask what they are free to share about potential openings and their hiring procedures.

Develop at least one style of résumé and keep it up to date. Strive to provide an honest assessment of your skills and qualifications, without embellishing your accomplishments. Seek the input of respected peers to help you accurately evaluate your qualifications. Consider creating résumés in a couple different formats so that you can cater the content to highlight the most relevant aspects of your background for a particular job.

Scrutinize your contact information, automatic replies, voicemail greeting, and email address. Are they professional or could they be perceived as too casual, or worse, offensive or suggestive? Likewise, have you posted anything on social media that would be perceived as offensive or suggestive? Given equally qualified candidates, employers will likely select the candidate who presents less risk of being offensive or inappropriate on the job.

Interviewing

Be prepared. Reread the job description very closely and research your potential employer. Anticipate what questions will be asked of you during the interview and practice your responses. There are several standard questions you are likely to encounter, for example—"Tell me a bit about yourself and why you are a good candidate for this job and this agency." This question is designed to get you talking and to determine if you understand the qualifications for the job. Another common question is "How will you be of benefit to this agency?" This question is designed to determine if you did research on the nature of the employer and its mission. There are several websites with lists of standard inter-

Practical Matters 101

view questions that you can consult. Try to anticipate what questions might be asked about specific job skills. Don't be surprised if asked to solve math problems specific to a job, or for a handwriting sample if the position involves data collection.

Interviews should be two-way exchanges of information, with a chance for the candidate to learn more about the potential employer and the nature of the job. Candidates are often given the opportunity to ask questions about the job. This is not the time to ask about or negotiate pay. Rather, ask a question that demonstrates your interest in the position and your desire to add value to the agency or organization. Again, there are many websites with lists of possible questions to ask of your potential employer. Modify these to formulate questions relevant to the particular job and agency to which you are applying.

Keep in mind that you may be observed both before and after your interview. Some bosses will wait in the lobby for an interview candidate and observe them enter and greet the receptionist. An applicant who is rude or dismissive to a receptionist might be ruled out even before the formal interview. The logic is that employees represent the employer, and if a candidate cannot be polite and well behaved when on their "best behavior" for the interview, he or she will likely be a poor representative of the company once hired. Do not smoke, chew gum, or chew tobacco during an interview. Remember to thank everyone who participated in the interview. Even if you do not get the job, consider the interview a chance to practice and polish your interviewing skills.

When deciding what to wear for an interview, avoid immodest clothing or anything with slogans or logos. Dress a step up from the clothes you would be expected to wear at the job for which you are applying. So, if you are applying for a field job with an interview in an office, wear business casual attire. If the job would normally entail wearing business casual, then wear formal business attire. If the interview takes place in the field, wear neat and clean field clothes. It is seldom appropriate to wear a formal suit to an interview for a beginning field job. Conversely, do not wear field clothes to an interview for an office job with the expectation of being hired entirely due to your skills.

Providing References

Ask the people you wish to use as references what type of reference they will provide for you before listing them on an application. Ask them to be candid about their opinion of you and what they are willing to tell a prospective employer. Some people may not be willing to serve as a reference, or have constraints from their employers about what they can say about you. Some references may require you to give them written permission to speak to a prospective employer. There are two key questions that employers will likely ask your references. The first question is whether the person is, in fact, willing to provide information and serve as a reference. If the person you listed will not provide a reference for you, this is an indication to the employer that you did not check with that person before listing him or her on your application, and the person is unwilling to say something bad about you. Your chances of being hired will be slim if this happens. The second key question is whether the person providing the reference would hire you for the job. If your reference claims that he or she would not hire you for a job, you will likely not be hired. So, check with your references first and ask them to candidly tell you how they will answer those questions.

It is also good practice to provide a current résumé to those you list as references. Inform them when you schedule interviews so they are prepared for a reference check rather than being surprised by an unexpected request. Additionally, remember that the community of natural resources professionals is small. Your prospective employer may know someone at the school you attended or at your current, or previous, workplace. Depending on their organization's policies, your former classmates, coworkers, or instructors may be free to talk about you to a prospective employer.

Tips from a Career Guide

The following tips have been adapted from *A How-To Guide for Pursuing a Career in Natural Resources* (Colorado Alliance for Environmental Education 2018).

1. Flexibility is very important, especially early in your career. You may need to move to a different part of your state or even a different region of the country to get an entry-level job to gain the experience you need to reach your long-term goals. Similarly, you may need to take an entry-level position that you aren't particularly passionate about just to get your foot in the door at an agency where you eventually want to work as a professional. Consider the potential of the experience you gain on the job rather than the pay or glory of the job.

2. If you are a current college student, or a very recent graduate, apply to the Pathways Program. This is a federal government program specifically designed to help guide college students into careers with the federal government in various agencies. You have a better chance at getting into a federal career through the Pathways Program than through the regular competitive process. So, take advantage of this opportunity while you can!

3. One of the most important things you can do for your career is build relationships and network with others in your profession. No matter what your personality, be bold! Step outside of

your comfort zone and meet new people in positions and at organizations in which you are interested. Volunteer, join a professional society, attend job fairs, reconnect with your teachers and classmates, and visit local conservation agencies' offices.

4. Develop good communication and analytical skills. Few natural resource jobs involve working alone in the middle of the forest on simple tasks. Most jobs, especially as you gain experience, require thinking about the tasks assigned. And you will need to work closely with others—including landowners, politicians, partners, and coworkers. Build good written and verbal communication skills, critical thinking skills, problem-solving skills, and mathematical skills. Be sure to list these skills on your résumé.

5. It's easier to get a job if you already have one. So be open to part-time positions and internships that will fill in the time on your résumé and keep you in contact with people and issues in the natural resources field. In many agencies, seasonal positions are the best way to be considered for a permanent position, so if you do not get a full-time permanent job right out of college, be prepared to accept a part-time or seasonal position. It may be better to accept a low-paying job that provides the experience you desire and for which you have a passion rather than a high-paying job that is uninteresting and may not lead in the direction you wish to pursue.

6. Strive to obtain the job skills that you don't already possess. Read position descriptions that interest you and if you don't feel qualified or keep submitting unsuccessful applications, find the gaps in your qualifications and fill those through volunteer work, self-study, taking classes, obtaining a certificate, or even obtaining another degree.

7. Your résumé for a government job should look different from your résumé for a private or nonprofit position. For any job application, you want to tailor your résumé to the duties and qualifications listed in the position announcement. Government agencies have strict requirements for meeting the qualifications listed on their job announcements. For example, if the job requires one year of specialized experience in counting birds, you need to include on your résumé at least one full year of experience counting birds. This means 40 hours a week for 52 weeks, not part time for 18 months, or full time for 51 weeks. Read each qualification carefully and support it specifically with experience on your résumé. Résumés for private industry positions should be short and to the point. Résumés for federal government jobs should be longer and include key words about your skills and previous job tasks and duties.

8. Recent college graduates seeking jobs with the federal government should look for GS-4 positions if they have an associate's degree and GS-5 positions if they have a bachelor's degree. Many youth and entry-level candidates are unsure of which positions they qualify for. Government agencies use a code called a "General Schedule" or "GS" code for the level of their professional positions based on education and experience. New graduates with no experience should realistically look at position announcements with a GS-4 or GS-5 code depending on their degree.
9. If you are a veteran or have a disability, your chances of getting a job with the federal government are very good. The federal government employs the greatest percentage of veterans and persons with disabilities since it desires to help them get good jobs and make a sustainable living. As a result, there are many government programs to help veterans and people with disabilities obtain federal employment.
10. Find an agency that fits your passion. Natural resource agencies have varying missions and organizational cultures. Take some time to explore the jobs, people, and places associated with each agency, and focus on getting a job at an agency that most closely matches your own ethics and interests.

Bonus Tip: You can make a living doing what you love. Depending on your education, experience, and the career you choose, natural resources professionals make anywhere from $22,000 to more than $140,000 a year. So be patient, persistent, and flexible in developing your career.

Human Dimension Skills

Employers are looking for workers who possess both technical know-how and people skills; in other words, society-ready individuals (see Chapter 5).

Communication

Proficiency in both oral and written communication is important. Be aware of the audience of your communication and speak with the proper formality when appropriate. Note that written communications are nearly always more formal than oral communications. Do not write like you speak. In natural resources, most of the communication will be technical in nature, either business or scientific. Learn to write using both styles.

Practical Matters 105

Natural resources professionals need communication skills for a variety of audiences. Here, Jim Pena, Regional Forester for Region 6 of the Forest Service, addresses a Job Corps graduating class. (Photo by USFS, Flickr)

Use proper telephone and email etiquette. When sending emails, use the subject line to briefly describe the email's content. Most busy people and those aware of email security will not open an email with no subject line. Write, don't text; use capitalization and punctuation and avoid trendy abbreviations. Use the appropriate level of formality needed for the communication.

Ethics

Each of the professional societies mentioned in this text has a code of professional ethics or a code of conduct, and all are available on the Internet. Familiarize yourself with these ethics codes in order to understand what conduct is considered ethical in the context of the natural resource professions. Private companies and government agencies will have ethical standards as well. For example, Weyerhaeuser's code of ethics is also available on the Internet. The US Department of Agriculture has an ethics website with information on federal ethical standards and specific examples for the Forest Service and its employees. It even has an interactive self-help guide to assist employees with ethical questions.

Basic and Professional Etiquette

Be Courteous. Introduce other people and use the proper titles for those you address and introduce. Realize that some college programs and/or agencies operate on a first-name basis, while others are more formal, with instructors and supervisors who prefer to be addressed by title. In some circumstances, such as when performing fieldwork, it may be appropriate to address your boss, or professor, by first name, but once back in the office or classroom, it is appropriate to be more formal and use titles (especially if higher level supervisors or college officials are present). It is always appropriate to start out using formal titles and to do so until asked to address someone by his or her first name. Politeness is always appropriate in the workplace. Send a formal thank you note whenever you are given a gift, attend a party, or taken out for a meal. Learn and use appropriate table manners.

Appearance and personal grooming do indeed matter. Dress appropriately in neat, clean, serviceable clothing suitable for the tasks of the day, whether in the field or in the office. Dress nicely if you are expected to meet the public. Do not assume other offices have the same dress code as yours.

Learn to Travel Efficiently. Depending on the natural resource program you chose, you will likely take field trips, perhaps overnight field trips, and be encouraged to attend multiday professional seminars and conferences. You may be expected to travel for training or for remote work assignments. Learn to pack appropriately and efficiently, making sure you have the correct gear and are prepared for weather changes and impromptu social engagements—without taking more space than necessary. Firefighters, for example, usually have a weight and size limit for what can be carried. But even when traveling with coworkers, regardless of the purpose, pack efficiently.

Working in Groups

Recognize that you will encounter people from different generations and from different regions of the country, all with backgrounds that may be dissimilar to yours. Whether in college or in the workplace, be respectful of others. Realize that you will be working with people who are passionate about the resource, but that passion may be expressed in different forms. For example, many natural resources professionals hunt and fish, many others do not; some choose not to spend their free time in wildlands. Respect people's differences.

Teamwork is common in natural resource work. Learn to collaborate effectively with others. Share the work load and give credit to others when appropriate. Others may be less skilled than you, or may possess skills that you do not. Learn from them, and teach them the skills you have. Accept that others have different physical, intellectual, and emotional limits than you.

Records and Accounting

Increase Your Field Note-Taking Skills. Despite the increased use of tablets and data loggers, handwritten notes and sketched maps and diagrams are still important and often the most appropriate method of

Practical Matters 107

Fisheries biologists, working as a team, electroshock fish as part of a stream survey project. (USFS Colville National Forest, Flickr)

recording data in the field. Learn to write and print so others can read your writing. Consider carrying a pad of all-weather paper and either an all-weather pen or hard lead pencil. Soft lead pencil will smear.

Educational Records. Keep your notes and materials, especially from your advanced courses. It is likely you will need to refer to these later in your education or career.

Employment Records. Keep records for each of your jobs. This should include your exact start and end dates; job title; major assignments and tasks; pay rate and grade; and supervisor's name, title, and contact information. It is also a good idea to keep a brief journal of your daily tasks. All of these records will help you write future résumés and job applications quickly and accurately.

Training, Continuing Education, and Volunteer Experience Records. Maintain records of any training, certifications, continuing education, and volunteer experiences.

Bookkeeping and Accounting. Learn some basic accounting skills, financial vocabulary, and financial concepts such as the time cost of money (interest) and the difference between gross and net income. Having a basic knowledge of bookkeeping and budgeting will be useful even in entry-level jobs.

Build a Personal Library of Reference Materials. Consider building a library of standard texts and references. Keep your course notes and textbooks from advanced courses within your discipline. Maintain your online reference library using bookmarks for frequently visited sites.

Natural Resources Field Skills

Local Ecosystem Characteristics

Natural resources professionals should be able to describe the characteristics of the ecosystems in which they work. This will include knowing the plants, animals and their habitat needs, common diseases, and insects that inhabit these ecosystems; the types and quality of soils present; and the area's geology. They should also be able to describe the role of fire and weather in these habitats, as well as the roles of the biotic communities.

An inventory crew samples forest understory. Early in their careers, natural resources professionals will likely spend a great deal of time collecting field data. (Photo by USFS, Flickr)

Practical Matters 109

Map reading and interpretation are important skills for natural resources professionals. Here, a ranger in the Wild Sky Wilderness in Washington State consults a map. (Photo by USFS Mt. Baker-Snoqualmie National Forest, Flickr)

Basic Land and Water Navigation

Learn how to stay found when traveling in wildland areas, as well as how to navigate efficiently and how to record accurate positions to document a location or return to an exact location. Learn how to measure distance using pacing. Pacing—using the distance between footfalls while walking—is a very common skill used in fieldwork and is accurate enough for quick measurements of distance. You should know how to use a magnetic hand compass and to correct for declination, and be proficient in using a GPS receiver both for navigation and measuring land area.

Outdoor Safety and Efficiency

First Aid Training. Most agencies and private firms require employees to have first aid training. If your employer doesn't require or offer first aid training, review training materials or obtain training on your own.

Appropriate Footwear. Acquire the appropriate footwear for a variety of seasons and weather. Preference for footwear seems to be regional and is changing with new technology and the cost of traditional style boots. Obviously, high heels, flip-flops, sandals, and open-toed shoes are not appropriate field wear. Non-slip, closed-toe footwear (generally boots) that protects the ankle is the minimum requirement. Boots must be appropriate to keep feet cool or warm and dry. The federal standard for wildland fire is "a minimum of 8-inch high, lace-type exterior leather work boots with Vibram-type, melt-resistant soles. The 8-inch height requirement is measured from the bottom of the heel to the top of the boot. Alaska is exempt from the Vibram-type sole requirement." Although they may be required by some regulatory agencies, steel-toed boots are not allowed in wildland fire fighting and are uncomfortable for hiking any distance.

Appropriate Clothing. Acquire the appropriate clothing for a variety of seasons, weather, and uses. Appropriate clothing will vary seasonally and regionally. Also, the clothing worn for outdoor recreation may not be suitable for outdoor work. Some natural resource jobs require both strenuous tasks and periods of sedentary work. Learn to dress efficiently for the desired task and level of activity with a margin of safety for cold or hot weather.

Consider obtaining a cruising vest. A cruising vest has many pockets to keep equipment accessible and prevent loss. Carrying equipment in pants pockets is inconvenient as the equipment will often work its way out of the pocket while you walk. Carrying equipment by hand also is inconvenient, and setting it on the ground between uses increases the chance it will get dirty, lost, or stepped on and damaged; it also slows down the work from having to reach down.

Natural resources professionals work in a variety of environments. (Photo by Ben VanAlen, USFS Sitka Ranger District, Tongass National Forest, Flickr)

Weather Safety. Lightning, flash floods, high winds, tornados, large hail, and extreme cold are all possible weather-related safety issues. Learn the appropriate means to mitigate the risk from these hazards while working in the field or on the water.

Water Safety and Basic Boating. Depending on the discipline, natural resources professionals may spend a lot of time on the water and in various watercraft. But any natural resources professional could find himself or herself on a small boat and should be aware of safe boating practices.

Animal Safety. There are many animal safety issues; some of the more obvious are encounters with potentially dangerous animals like bears, cougars, and moose, as well as sharks and large marine mammals. Snakes present a safety concern in some areas, as do many smaller animals along the ocean shore. Learn the appropriate precautions and proper behavior when working in an area with potentially dangerous animals.

Overhead Hazards. For those who work in the woods, overhead hazards are a safety issue. Learn to look up for snags and large branches hung up in trees.

Insect Safety. Wasps, bees, deer flies, horse flies, mosquitoes, chiggers, ticks, and other insects can be annoying as well as vectors of disease. Learn about the insects in your work area and how to prevent and treat their bites.

Firearms and Hunting Safety. Even those who do not use firearms professionally or for recreation should know some basic firearm and hunting safety practices. Anyone working in a wildland or rural area is likely to have at least casual contact with others who are target shooting, hunting, or just carrying firearms. If you observe what appears to be safe and responsible firearm use, you can judge whether it is safe to remain near an area where someone is hunting or shooting and, if they appear unsafe, how far to move away. For example, being a mile away from someone bird hunting with a shotgun is likely a safe distance; being a mile downrange of an irresponsible user of a high-powered rifle is far too close. Being aware of hunting rules, seasons, and practices will help you avoid times when it is unsafe, or discourteous, to enter an area used for hunting.

Rural Driving. There may be less traffic in rural areas but there are still many driving hazards. Loose gravel, washboard and out-sloped surfaces, soft shoulders and berms, narrow lanes or single lanes with turnouts, bridges without curbs or side rails, fords, free-range and wild animals, cattle guards, log trucks, and high water are common in rural driving. Learn the common hazards in your work area and how to avoid

Rural and wildland driving presents unique challenges. (Photo by author)

Practical Matters   113

or deal with them. Learn how to properly chock the wheels of a vehicle and determine if the vehicle needs to be protected from rodents when parked for long periods. Learn how to back up a vehicle fluidly and to back up a vehicle with a trailer. Learn to be a helpful spotter to assist, rather than hinder, the driver who is backing a vehicle. Unless directed otherwise, gates should always be left in the condition they are found, so reclose them if they were closed, and leave them open if they were open. It is courteous and efficient for the passenger to open and close gates as needed. Passengers can also assist the driver by watching for wildlife, road signs, and turn offs. On an unknown route, the passenger should track the vehicle progress on a map and inform the driver of the progress. When stopping for a break or refueling, the passenger should offer to clean the windows and, if qualified, check the fluids. One college program includes learning to back up a vehicle with a trailer as a required portion of its natural resource curriculum.

Defensive Driving. Most agencies and large organizations require employees to take a defensive driving course before driving their vehicles and to renew that course on a regular basis. Even if it is not a requirement, consider taking such a course on your own. Remember that a serious accident or even traffic citations while in company vehicles may affect your career.

Blocked Culverts. Some organizations have a policy that anyone driving on forest and range roads who observe a blocked culvert must stop and attempt to clear the culvert if it can be done safely, or report its condition immediately.

Manual Transmissions. Fewer vehicles are being equipped with manual transmissions, but some agencies may still have fleet vehicles with manual transmissions. If you have the opportunity, learn how to drive a vehicle with a manual transmission. Being able to drive all the vehicles in a fleet, or light tractors and other equipment, is an advantage.

Four-wheel Drive. If you have the opportunity, familiarize yourself with different four-wheel drive systems. Chances are you will be driving some type of four-wheel drive vehicle if you work in the field. There are a variety of different systems and it is a good idea to educate yourself on their differences and characteristics.

Basic Auto and Small Engine Maintenance and Repair. Working in the field will likely take you to remote areas where help will take some time to arrive even in an area with cell phone coverage. Learn to check the fluid levels, hoses, belts, lights, and tire pressure for a vehicle. Many organizations have a policy that it is the driver's responsibility to check these before driving. Also, having a sense of what is a go or no-go condition is beneficial; in other words, when do you stop driving and have the vehicle towed or fixed on site, or when can you drive safely for a distance without causing further damage. Chainsaws, outboard motors, pumps, snowmobiles, and four wheelers all use small engines. Some of these will be two-stroke engines; learn how to mix the oil and fuel for these engines and the basic maintenance and starting procedures for both two-stroke and four-stroke engines.

Heavy Equipment Safety. Chances are you will work around heavy equipment at some time in your career, perhaps often. Heavy equipment is used in restoration projects and timber harvesting. Realize that the operator of the equipment has large blind spots and the equipment may swing or turn very quickly. Never approach the equipment without eye contact with the operator, the operator's signal to approach, the blade or shovel dropped, and the equipment completely shut down with the engine turned off. If you are walking and about to be passed by a piece of heavy equipment or a truck, step off the road and stand where the equipment operator can see you until the equipment or truck passes. It is easy to lose your balance when walking next to a moving vehicle.

Radio Use. Learn the proper use of a radio following your organization's guidelines. Learn general radio procedures, such as listening first for a break before talking, and the specific operating procedures for your organization's radios. Check if there are more than one type of radio used in your organization and become familiar with all the types used. Do this before you need to use the radio in a stressful situation.

Units of Measure. Most likely you will be using both the metric and imperial system (US customary units) of measurement. Learn both and how to convert between the two systems.

Chainsaw Use. In forest regions these are commonly carried on vehicles to clear fallen trees on or near roads and trails, and for project work. Learn to operate a chainsaw safely and efficiently.

Hand Tools. Learn to safely use and sharpen basic hand tools such as a pocket knife, shovel, axe, Pulaski, and light saws.

Survival and Comfort. Pack an extra gallon of water and some food in your vehicle or boat for instances where you are delayed or unexpectedly work late. Many field workers carry a container of baby wipes and a toothbrush to freshen up. Another tip is to stuff a wad of toilet paper in a plastic bag and carry in your hard hat, work vest, or back pack. An extra pair of socks may come in handy after getting wet feet, or they can be used as mittens or ear muffs. A large bandana is also handy.

Practical Matters   115

Geographical Information Systems (GISs). Geographical information systems integrate mapping and data analysis. This has become a powerful tool for natural resource management and other disciplines. It will likely be required learning during your formal education or job training. Even if it is not required, learn about its use and application to your discipline.

Agency History. Learn the history of your agency or company, as well as that of the other agencies that affect yours. Knowing the history of your organization often helps you understand the origin of some policies and procedures.

SUMMARY

Practical matters are important to developing a successful career. Gaining basic experience, applying for jobs, and presenting yourself well during interviews are important beginning skills. Natural resources professionals also need to develop the human dimension skills of communication, etiquette, the ability to work with other people, and ethics. Working in the field is the part of the career that many professionals enjoy the most; but successful fieldwork requires the ability to work safely and accomplish the necessary tasks efficiently. Gaining and practicing these job-seeking, human dimension, and field skills will hopefully lead to a satisfying and productive career.

STUDENT EXERCISES

1. Go to the USAJOBS website and develop a profile.
2. Find an ad for a natural resource job that interests you. Write a résumé and cover letter as if applying for that job.
3. Use the Internet to find sites that list possible interview questions. Write down several questions that may be asked at an interview for a natural resources job.
4. Pick a safety hazard applicable to your region. Use the Internet to determine how to minimize the risk from this hazard.

WORKS CITED

Colorado Alliance for Environmental Education. 2018. *A How-To Guide for Pursuing a Career in Natural Resources. 2nd Edition*, Spring 2018. Accessed March 20, 2018. http://www.getoutdoorscolorado.org/career-guide-signup

Hutton Junior Fisheries Biology Program. n.d. Home page. Accessed September 17, 2017. https://hutton.fisheries.org/

Appendix A

Accreditation Standards

The Society of American Foresters accredits programs that award an associate degree in Forest Technology and programs that award a bachelor or master degree in Forestry, Urban Forestry, and Natural Resources and Ecosystem Management. One set of standards applies to Forest Technology programs, and there are some 24 schools in the United States and Canada with accredited Forest Technology programs (see the discussion of Forest Technology on pp. 124–125). Another set of standards applies to programs that award bachelor and master degrees, and there are 52 US schools with an SAF accredited bachelor or master program. Most of these accredited programs are Forestry programs as the SAF has only been accrediting Natural Resources and Ecosystem Management programs for a few years. It is important to note that accreditation is for individual programs and degrees—not a department, school, or college. In other words, a school may grant several degrees, but only a select few may be accredited. An example is the University of Washington School of Environmental and Forest Sciences, which has a bachelor's degree in Environmental Science and Resource Management that is not accredited, but the education gained through the degree will lead directly to candidate status as an SAF Certified Forester. In contrast, the University of Washington has an accredited Master of Forest Resources program. As another example, Colorado State University offers bachelor, master, and doctorate programs; but only the bachelor's program is accredited. The accredited programs are all listed on the SAF website.

The Society of American Foresters standards for curriculum are paraphrased here and were extracted from the SAF 2017 Accreditation Handbook (2016).

Society of American Foresters Standards for Forestry, Urban Forestry, and Natural Resources and Ecosystem Management (NREM)

The overall requirements of the SAF are that Forestry, Urban Forestry, and Natural Resources and Ecosystem Management (NREM) education must begin with the fundamentals: written and oral communication; mathematics; the biological, social, and physical sciences; and the humanities. The SAF standards for accreditation require a distinction between general education courses and those specific to professional forestry, urban forestry, or NREM. The professional courses need to provide in-depth coverage of biology, ecology, measurement of forest resources, management of forest resources, forest policy, forest economics, and administration. Students must take the professional curriculum sequentially, and the curriculum must be presented in a manner that fosters analytical and critical reasoning skills. These skills should include systematic problem solving and decision making, and should be grounded in an awareness of the current and historical issues and policies that affect forest resource management, urban forestry, ecosystem management, and conservation.

The SAF requires that the curriculum "provide a variety of educational experiences, including lectures, discussion, simulations, computer applications, and individual and group projects in laboratories and field experiences." The purpose of this variety is to ensure that graduates of an accredited program can develop, apply, and execute management plans that address owner's goals, forest health, forest sustainability, and both legal and regulatory requirements. Moreover, graduates will be able to apply scientific methodologies to attain a variety of sustainable forest or urban forestry products, services, and conditions, or natural resources and ecosystem services and conditions.

General Education

The general education requirements for the Forestry, Urban Forestry, and NREM programs are the same. These requirements are:

- Communications: oral and written communication skills must be developed throughout the curriculum. Students must have the ability to prepare and deliver effective oral presentations and a proficiency in English composition including technical writing, business writing, and writing for nonprofessional audiences. They must also be able to read with comprehension and understand, communicate, and critically evaluate multiple viewpoints.

- Biological Sciences: students must understand the components, patterns, and processes of biological and ecological systems across both spatial and temporal scales, understand molecular biology and possess an understanding of the structure and function of cells, organisms, populations, species, communities, and ecosystems.
- Physical Sciences: students must understand both physical and chemical properties, measurements, structure, and states of matter.
- Mathematics: students must apply mathematics and statistics for analysis and problem solving appropriate for their program's stated outcomes.
- Social Sciences and Humanities: the social sciences and humanities are critical components of a professional education. Students must demonstrate the ability to address moral and ethical questions and use critical-reasoning skills. They must possess an understanding of human behavior and an understanding of social and economic structures, processes, and institutions across a broad range of societies. They must also possess an understanding and appreciation for the diverse dimensions of the human experience and culture.
- Technological Literacy: students must be able to demonstrate the appropriate use of contemporary electronic technologies in their professional life.

Professional Education

The Professional Education portion of the SAF requirements become more specific and require the program to have depth, breadth, and balance among four major subject categories, with adequate instruction in basic principles, typical laboratory and field applications, and current practices in these four subjects.

Forestry Programs The specific requirements for Forestry programs are the following topics:

- Ecology and Biology: students must able to demonstrate an understanding of: (1) taxonomy and be able to identify forest and other tree species, describe their distribution, and describe the associated vegetation and wildlife of tree species. (2) soil properties and processes, basic hydrology, water quality, and watershed functions. (3) ecological concepts and principles, including the structure and function of ecosystems and that of plant and animal communities. (4) competition, diversity, population dynamics, succession, disturbance, and nutrient cycling. (5) demonstrate their ability to make ecosystem, forest, and stand assessments and their knowledge of tree physiology, and be able to explain the effects of climate, fire, pollutants, moisture, nutrients, genetics, insects and diseases on tree and forest health and productivity.

- Measurement of Forest Resources: students must demonstrate their ability to: (1) identify and measure land areas and conduct spatial analysis. (2) design and implement resource inventories that meet objectives using appropriate sampling methods and units of measurement. (3) analyze these resource inventories and project future conditions for forests, stands, and trees.
- Management of Forest Resources: students must demonstrate their ability to: (1) develop, apply, and understand the effects of silvicultural prescriptions that are appropriate to management objectives, including an understanding of the methods of establishing and influencing the composition, growth, and quality of forests. (2) analyze the economic, environmental, and social consequences of forest resource management strategies and decisions. (3) develop management plans with multiple objectives and constraints and demonstrate their understanding of the valuation procedures, market forces, processing systems, transportation, and harvesting activities that transform human demands for timber-based and other consumable forest products into the availability of those products. (4) demonstrate an understanding of the valuation procedures, market, and non-market forces that provide humans the opportunities to enjoy nonconsumptive forest products and services. Students must also demonstrate an understanding of the administration, ownership, and organization of forest management endeavors.
- Forest Resource Policy, Economics, and Administration: students must demonstrate an understanding of (1) forest policy and the processes by which it is developed and demonstrate an understanding of how federal, state, and local laws and regulations govern the practice of forest resource management. (2) professional ethics, including the SAF Code, and a recognition of the responsibility to adhere to ethical standards in decision making on behalf of both clients and the public. (3) the financial, human resources, and legal aspects of public and private enterprises.

Urban Forestry Programs The specific requirements for Urban Forestry programs are the following topics:

- Ecology and Biology: students must demonstrate an understanding of: (1) taxonomy and an ability to identify a wide range of woody plant species, including both native forest species

Accreditation Standards

and introduced urban forest species and these tree species' growth and health in urban and suburban environments. (2) soil properties, biology, and processes, especially soil nutrients, soil compaction issues and mitigation, basic hydrology, water quality, and watershed function. (3) ecological concepts and principles including the structure and function of ecosystems and especially the growth and performance of various tree species in urban and suburban settings, plant and animal communities common to urban forests, diversity, and disturbance. (4) arboriculture including an understanding of tree establishment and maintenance in urban settings, and the use of basic tools and equipment in arboriculture. (5) tree physiology and anatomy and the effects of climate, fire, pollutants, moisture, genetics, construction, insects, diseases and cultural problems on tree health and urban forest composition. (6) urban wildlife and its interactions with urban forests, and a thorough understanding of decay and defect development in individual trees, tree failure patterns, and how to conduct a tree structure evaluation. (7) the practice of arboriculture and urban forestry in multiple cultural, physical, and population densities.

- Measurement of Urban Forests and other Forest Resources in Urban Settings: students must demonstrate the ability to: (1) identify and place trees in appropriate locations relative to other trees and relative to structures in an urban and suburban environment. (2) evaluate the health and benefits of urban forests; be able to measure, assess the appropriateness of existing trees and urban forests; be able to assess tree risk and health. (3) apply statistical sampling methods and possess the ability to conduct and analyze inventory data to model future urban forest changes, assess green space, and monitor tree health. (4) conduct spatial analysis and utilize GIS and remote sensing tools/skills in urban-rural interfaces. (5) apply appropriate appraisal methods to value urban trees considering species, site, landscape location, condition, and market value.

- Management of Urban Forests and Trees in an Urban Environment: students must demonstrate an ability to: (1) develop and apply prescriptions (plans) appropriate to management objectives, including methods of establishing and influencing the species composition, growth, and quality of trees and urban forests and to understand the impact of those prescriptions. (2) analyze the economic, environmental, and social consequences of urban forest management strategies and decisions and an ability to develop management plans with specific multiple objectives and constraints. (3) demonstrate a knowledge of urban and land use planning including the fundamentals of site planning and landscape design. (4) demonstrate a knowledge of basic accounting and business skills, including budget development and management. (5) demonstrate an understanding of wildlife habitat management in urban environments, including habitat enhancement and wildlife pest management.

- Urban Forest Resource Policy, Economics, and Administration and Law: students must demonstrate: (1) an understanding of the processes and legal aspects of municipal policy. (2) an understanding of federal, state, and local laws, ordinances, and regulations relative to the practice of urban forestry. (3) an understanding of professional ethics and recognition of the responsibilities to adhere to ethical standards in urban forestry decision making on behalf of clients and the public. (4) an ability to communicate as an urban forestry professional; an understanding of industry best management practices and the applicable federal, state, and local guidelines and standards for safety and performance. (5) an understanding of the administration, ownership, and organizational implications for urban forests under both private and public ownership, including appraisals.

Natural Resources and Ecosystem Management (NREM) Programs
The specific requirements for NREM programs are a fundamental knowledge of the following topics.

- Ecosystem Components and Ecosystem Function, including Human Systems: students must demonstrate: (1) a knowledge of the elements of botany, zoology, entomology, plant pathology, plant physiology, and genetics essential to an understanding of higher-order ecological processes and an understanding of taxonomy and systematics and an ability to identify dominant and/or ecologically significant components of the flora and fauna of ecosystems at regional to continental scales. (2) a knowledge of the important life history characteristics of dominant and special-concern species. (3) a knowledge of soil properties and processes, basic hydrology, water quality, and watershed functions. (4) an understanding of ecological concepts and principles including the structure and function of ecosystems, plant and animal communities, competition, diversity, population dynamics, succession, disturbance, and nutrient cycling. (5) an understanding of the effects of climate, fire, pollutants, moisture, nutrients, insects and diseases, and other environmental factors on ecosystem health and functioning at local and landscape scales.
- Measurement and Assessment of Ecosystem Components, Properties, and Functioning, including Human Systems: students

must demonstrate an ability to: (1) identify, measure, and map land areas and conduct spatial analyses, and an ability to design and implement accurate inventories and assessments of dominant or critical ecosystem components and services, ecosystem properties, and indicators of ecosystem health, including trees and other vegetation, vertebrate fauna, biodiversity, soil and water resources, timber, and recreational opportunities. (2) summarize and statistically analyze inventory and assessment data, evaluate the status of important ecosystem components, describe and interpret interactions and relationships, and project future ecosystem conditions.

- Identification and Evaluation of Management Objectives: students must demonstrate: (1) an understanding of the valuation procedures, including market and non-market forces, that apply to ecosystem goods and services such as timber, water, recreational opportunities, carbon and nutrient cycling, and plant and animal biodiversity. (2) an ability to explain the relationships between the demand, cost of production, and availability of those goods and services and an ability to describe procedures for measuring stakeholder values and managing conflicts in the evaluation and establishment of management objectives. (3) an ability to evaluate and understand the economic, ecological, and social trade-offs of alternative land uses and ecosystem management decisions at local, regional, and global scales, and a knowledge and understanding of environmental policy as applied to ecosystems and the processes by which policy is developed.

- Management Planning, Practice, and Implementation: students must demonstrate: (1) an ability to develop and apply prescriptions for manipulating the composition, structure, and function of ecosystems to achieve management objectives, and to understand the impacts of those prescriptions at local and landscape scales. (2) an ability to identify and control or mitigate specific threats to ecosystems such as insects, diseases, fire, pollutant stressors, and invasive plants or animals. (3) knowledge of the methods and procedures unique to the production of ecosystem goods and services such as timber, recreation, water, and wildlife populations and the ability to describe the process of adaptive management and its application to the management of ecosystems. (4) an understanding of how federal, state, and local laws and regulations apply to management practice. (5) an ability to develop management plans with specific objectives and constraints that are responsive to ownership or stakeholder goals and demonstrate clear and feasible linkages between current condition and desired future condition. (6) an understanding of professional ethics, including the SAF Code, and recognition of the responsibility to adhere to ethical standards in the practice of natural resource management on behalf of clients and the public; and an ability to integrate the knowledge, understanding, and skills from prior coursework in the development of collaborative solutions to realistic management problems.

SOCIETY OF AMERICAN FORESTERS STANDARDS FOR FOREST TECHNOLOGY PROGRAMS

The requirements for Forest Technology programs are quite different than those for Forestry, Urban Forestry, and NREM. Forest Technology programs must provide instruction in both oral and written communication, mathematics, natural and physical sciences, and social sciences. The forest technology curriculum must impart student proficiency in the following technical subject areas with a depth of instruction that reflects regional priorities and practices. The following topics are to be emphasized.

- Dendrology: students must be able to: (1) identify regionally important species by characteristics of leaves, twigs, bark, and fruit; know the family, genus, and species of important trees. (2) describe the regionally important species associations and succession. (3) must know the major commercial species of trees in North America and their uses. (4) be able to use dichotomous keys.
- Forest Ecology: students must be able to define and explain the importance of plant succession, site, soils, silvics; environmental protection, weather and climate influences, relationships of trees to other organisms, biodiversity, and ecosystems.
- Silviculture: students must be able to describe methods of regeneration, site preparation, planting practices, intermediate treatments, nursery and seed orchard practices, pesticide use and application, prescribed burning, pre-commercial thinning, commercial thinning, and harvesting systems.
- Protection: students must be able to define and describe fire management; threats to forest health; and regional insect, disease, and animal damage problems and control.
- Measurements: students must achieve competence in forest measurement and measurement equipment, log scaling practices, forest product measurement, sampling statistics, cruising and

Accreditation Standards

inventory techniques, log rules and volume tables, log and tree grading, growth measurement, and computer use and applications.

- Land Surveying: students must be able to use a hand compass, surveying equipment, pacing, tapes; read conventional and computer maps; conduct deed and title searches; understand land descriptions; use global positioning systems (GPS), and use geographic information systems (GIS).
- Remote Sensing: students must be able to acquire and process primary data derived from various sensors; identify and interpret remotely sensed data; determine bearings, distances, area, scale, and height; locate roads; and conduct type mapping.
- Woods Safety: students must know basic first aid; be able to identify hazards; understand hand and power tool safety; and use pesticides safely.
- Forest Products Manufacturing: students must understand the importance of regionally manufactured primary and secondary forest products, and the manufacturing processes by which forest products are produced.
- Harvesting Techniques: students must understand harvesting systems, cost analysis, logging plans, wood identification, road layout and construction, and best management practices (BMPs).
- Multiple Use of Forestland: students must demonstrate an understanding of forest wildlife, fish habitat, recreation, wilderness, watershed, timber, range, minerals, public conflicts, and public participation.
- Forest Management Practices: students must demonstrate an understanding of timber appraisal, contracts, forest management and ecosystem management principles, regional forest management regulations, sustainable forest management concepts and certification, and business recordkeeping and basic accounting.
- Human Resources Management: students must demonstrate an understanding of the behavior of groups and individuals, motivation, leadership, team building and dynamics, planning, decision making, rating and evaluation, workforce management, conflict resolution, and ethics.
- Students must complete a forestry-related work experience of at least 80 hours; this can be fulfilled through employment or a comprehensive field project. The experience should simulate working conditions of typical employers and include full-day schedules with appropriate assignments.

The Society for Range Management Accreditation Standards

The Society for Range Management (SRM) also accredits programs. There are 11 university programs in Rangeland Ecology and Management (REM) accredited by the society in the United States and one in Mexico. The society requires that an accredited REM program include the following topics:

- General Concepts: Biology (4 credits), Chemistry (4 credits), Soil Science (4 credits), and Plant Taxonomy (3 credits) that should include elements of both sight identification, plant classification, and keying.
- Quantitative Concepts (9 credits) with courses in Mathematics (college algebra or higher), Statistical Methods, Geographic Information Systems (GIS), Remote Sensing, or Natural Resource Modeling.
- Integrated Natural Sciences (9 credits) with courses in Ecology, Plant Physiology, Animal Physiology/Nutrition/Behavior, Biogeochemistry/Environmental Chemistry, Soil Genesis and Classification, Conservation Biology, Hydrology or Environmental Chemistry.
- Resource Management (9 credits) with courses in Watershed Management, Forestry Management, Wildlife Management, Wildland Recreation Management, Farm/Ranch Management, Fire Management, or Integrated Pest Management.
- Economics (3 credits).
- Communication (3 credits) with courses in Speaking, Writing, and Listening.

Rangeland Ecology and Management (REM) Specific Concepts

In addition to the general concepts listed above, 18 credits in REM specific concepts are required and expected to build upon the general concept listed above. The following specific subjects should be covered:

- Applied Rangeland Ecology: This topic should develop an appreciation of the full spectrum of considerations necessary to possess an awareness of the structure and functional dynamics of rangelands, the ability to develop healthy rangeland communities, and maintenance of healthy rangelands.

- Inventory and Assessment Methods, which should include instruction in quantitative and qualitative assessment of plant communities, land management units, the application of spatial analytical skills (such as mapping and the use of GPS, GIS, and remote sensing), and the application of mathematics and statistics.
- Vegetation and Habitat Management Techniques, which should include instruction to provide a "toolbox" of scientific methods that can be used to craft solutions to unique challenges. Examples of the challenges are fire and grazing management, restoration practices, weed management, watershed management, and riparian management.
- Rangeland Management Planning and Problem Solving, which should include elements of team projects and assessment of the ability to solve natural resource problems considering ecological, social, governmental policy, and economic contexts while using inquiry, analytical, integrative, and communication skills.

Instruction in the Rangeland Ecology and Management specific coursework should cultivate ethics, professionalism, and consideration for relevant environmental laws and policy. This would include an awareness of the rangeland management implications of the National Environmental Policy Act, the Endangered Species Act, the Multiple Use and Sustained Yield Act, and an appreciation of human-dimension considerations within rangeland management. Examples of the desired human-dimension skills for range managers would be negotiation skills and the ability to foster collaboration. The Rangeland Ecology and Management specific coursework should also strengthen technical writing skills.

WORKS CITED

Society of American Foresters. 2016. *Accreditation Handbook 2017*. Available as a PDF download at www.eforester.org.

Society for Range Management. 2015. *SRM Accreditation Handbook*. Available as a PDF download at rangelands.org.

Certification Requirements

Certifications are for individuals, whereas accreditation is for an entire college program. What follows are summaries of the requirements to become a Certified Forester by the Society of American Foresters, a Certified Wildlife Biologist by The Wildlife Society, a certified Fisheries Professional by the American Fisheries Society, or a Certified Professional in Rangeland Management by the Society for Range Management. The focus of these summaries is on formal education and work experience. Certification as a Forester or a Professional Range Manager also requires applicants to pass an examination; the details of the examination process are not provided here.

SOCIETY OF AMERICAN FORESTERS: CERTIFIED FORESTER

The Society of American Foresters requires the appropriate education, a minimum of five years of appropriate work experience, and the passage of an examination. Certified foresters must participate in 60 hours of continuing education during each three-year term of certification.

Education

There are two methods of obtaining the qualifying education (Society of American Foresters 2017a). Educational option one requires an earned bachelor's or master's degree from an SAF-accredited degree program. Educational option two requires a bachelor's degree with a minimum of 56 credit hours, of which 51 credits are in the following subject matter, with an additional 5 credit hours in any forestry-related coursework area (Society of American foresters 2017b). The four subject matter areas are:

- Ecology and Forest Biology with a minimum of 15 semester credit hours of dendrology, forest vegetation, plant taxonomy, forest soils or advanced soils at the junior level or higher, forest ecology or biology, silvics, wildlife biology or ecology, conservation biology, wetland ecology, forest entomology and pathology, integrated pest management, fire ecology, or wildfire management.
- Measurement of Forest Resources with a minimum of 12 semester credit hours of forest mensuration, forest resource measurements, land surveying, photogrammetry or remote sensing, principles and applications of geographic information systems, forest resource inventory, forest inventory design, sampling methods and analysis, or statistics.
- Management of Forest Resources with a minimum of 15 semester credit hours of silviculture, advanced forest ecology at the senior level or higher, forest resource management plans including the design, development, implementation and analysis of timber, watershed, wildlife, recreation, endangered species or urban forest applications, forest engineering, forest operations, timber harvesting, or other forest resource harvesting.
- Forest Resource Policy, Economics and Administration with a minimum of 9 semester credit hours of forest/natural resource policy including policy development and administration or the development and implementation law and regulation, forest or natural resource economics, markets, human resources, finance, business management, professional ethics, responsibility and integrity, standards of practice, or client/public/professional relationships.

Professional Experience

Certified Forester applicants must have a total of five years or more of qualifying professional forestry experience within the past 10 years in two of the four experience areas listed below (Society of American Foresters 2017c). Candidate Certified Forester applicants are those who have less than five years of qualifying professional forestry experience within the past 10 years.

Resource Assessment This category includes these activities: (1) Collecting preliminary data for a parcel of forest land, for example soils, cover types, access, stream and riparian areas.(2) Inventorying selected resources using accepted quantitative and/or qualitative methods. (3) Inventorying forest conditions such as weeds, insect, disease surveys, fuel loading, and damage using accepted survey methods. (4) Delineating property boundaries using appropriate methods and licensed surveyors when required. (5) Performing a resource supply-and-demand assessment for a discrete geographical area to determine availability and market conditions. (6) Determining the potential productivity of the land base for identified resources using accepted procedures to evaluate management options.

Management Planning This category includes these activities: (1) Confirming land ownership using legal records to assure the authority to implement management decisions. (2) Determining management goals, using stakeholder analysis, that establish priorities and direction for management. (3) Using resource assessment to describe existing resource conditions to provide a basis for developing science-based management options. (4) Developing management options by evaluating economic and operational factors to meet owners' objectives. (5) Establishing management options that meet landowner objectives and address foreseeable conflicts by using stakeholder input and resource assessment. (6) Developing compliance strategies by identifying applicable standards, regulations, and practices and by reviewing appropriate federal, state, and local laws, regulations, and voluntary practices. (7) Formulating the silvicultural system and associated practices as appropriate to achieve the established owner objectives. (8) Establishing monitoring and adjustment strategies to ensure that owner objectives are met and conflicts mitigated.

Stakeholder Analysis and Relations This category includes these activities: (1) Identifying potential stakeholders and discerning the level of their involvement in developing a strategy or management plan. (2) Evaluating the relative importance of each stakeholder's position to determine their level of impact on management planning and implementation. (3) Soliciting input as appropriate by engaging stakeholders and incorporating their concerns effectively in management planning and implementation. (4) Assisting landowners in establishing objectives by reviewing management options and their implications. (5) Advocating the importance of science-based forest policies, laws, and practices using appropriate channels of communication and influence to ensure the long-term capacity of the land to provide the variety of goods and services required by society.

Execution of Management Plans This category includes these activities: (1) Implementing management plans to meet owner objectives using specified activities such as surveying, harvesting, reforestation, site preparation, hazard reduction, and road building and in compliance with applicable laws, regulations, and voluntary practice standards. (2) Developing a budget by estimating costs and revenues for specified activities to fund the management plan. (3) Preparing contracts or work plans by developing and negotiating detailed specifications to implement the management plans. (4) Administering contracts or work plans that ensure monitoring and enforcement to meet management plan objectives. (5) Monitoring activities by measuring specified variables and indicators to ensure that the goals of a management plan are met. (6) Identifying changes as they occur by monitoring indicators with the intent to adapt management plans.

Continuing Education

Each Certified Forester must accumulate 60 Continuing Forestry Education (CFE) credit hours within a three-year period. At least 40 of these hours must be in Core Education (topics in Resource Assessment, Management Planning, Stakeholder Analysis and Relations, or Execution of Management Plan) and no more than 20 hours in Related Education (such as computer science, wildlife, or fisheries topics) or Professional Development and Volunteer Activities (such as writing articles or serving on boards). Candidate Certified Foresters must complete 15 hours of CFE annually with at least 10 hours in the Core Education category.

THE WILDLIFE SOCIETY: CERTIFIED WILDLIFE BIOLOGIST

The Wildlife Society's Certified Wildlife Biologist requires that applicants have, at a minimum, completed a course of study in a college or university leading to a Bachelor of Science, Bachelor of Arts, or the equivalent. At least one degree must have been completed in a wildlife-related field. An applicant must also have at least five years of applicable work experience. Applicants who do not meet the specified minimum educational requirements but who have more than 20 years of professional wildlife experience may qualify. Certified Wildlife Biologists must obtain a minimum of 80 hours of participation in organized activities or mentorship over a five-year recertification period.

Education

The applicant must have 36 semester credit hours from the first five subcategories listed below. Note that the sum of the hours required from those subcategories is 33 semester hours. The remaining three hours may be in any of these five subject areas. Course credits may be divided, but not duplicated, among categories when a course covers material in more than one category. High School Advanced Placement courses may be accepted. Comparable experience may be substituted for educational experience; however, applicants must have at least one college or university course in each category.

Certification Requirements

- Wildlife Management, a minimum of 6 semester credit hours in courses emphasizing the principles and practices of wildlife management. Courses should focus on understanding and manipulating wildlife habitats and population dynamics in the context of human objectives and influences. Conservation biology courses count if they contain a specific focus on management and decision making.
- Wildlife Biology, a minimum of 6 semester credit hours in courses in the biology and behavior of birds, mammals, reptiles, or amphibians. Courses should focus on the biology of wildlife species and their habitat relationships as the basis for management, and must include at least one course dealing solely with the science of mammalogy, ornithology, or herpetology. This course must be taken at a college or university and cannot be substituted by another course or experience. Ichthyology, marine biology (except courses about marine mammals or reptiles), microbiology, entomology, or related courses will not count in this category, but will qualify in the Zoology category.
- Ecology, a minimum 3 semester credit hours in courses in general plant or animal ecology excluding human ecology.
- Zoology, a minimum 9 semester credit hours in courses in taxonomy, biology, behavior, physiology, anatomy, and natural history of vertebrates and invertebrates. Courses in genetics, nutrition, physiology, disease, and other biology or general zoology courses are accepted as are courses in ichthyology or fisheries biology. Credits in general genetics and general biology should be split evenly between the Zoology and Botany categories.
- Botany, a minimum 9 semester credit hours in courses in general botany, plant anatomy, plant genetics, plant morphology, plant taxonomy, plant physiology, and other botany courses. Only one of the following courses is accepted: dendrology, silvics, or silviculture. At least one course must be primarily concerned with plant taxonomy or identification, and this course must be taken at a college/university and cannot be substituted by another course or experience. Credits in general genetics and general biology should be split evenly between the Zoology and Botany categories.
- Physical Sciences, a minimum 9 semester credit hours in courses such as chemistry, physics, geology, or soils, with at least two disciplines represented.
- Basic Statistics, a minimum 3 semester credit hours of course(s) in basic statistics.
- Quantitative Sciences, a minimum 6 semester credit hours in calculus, biometry, college algebra, advanced algebra,

trigonometry, systems analysis, mathematical modeling, sampling, computer science, or other quantitative science. Elementary algebra, introductory algebra, algebra, remedial algebra, introductory GIS, and introductory personal computing classes do not count.
- Humanities and Social Sciences, a minimum 9 semester credit hours in courses such as economics, sociology, psychology, political science, government, history, literature, or foreign language.
- Communications, a minimum 12 semester credit hours in courses designed to improve communication skills such as English composition, technical writing, journalism, public speaking, or use of mass media. A maximum of three semester hours each will be allowed for a completed Master's thesis and PhD dissertation. Courses in literature interpretation, foreign languages, and classes requiring a term paper, class projects, and seminars in non-communication courses will not count toward this category.
- Policy, Administration, and Law, a minimum 6 credit hours in courses that demonstrate significant content or focus on natural resource policy and/or administration, wildlife or environmental law, or natural resource/land-use planning will apply, in addition to courses that document contributions to the understanding of social, political, and ethical decisions for wildlife and natural resources management. Up to three semester hours in classes dealing with human dimension issues may count in this category, depending on course content. Conservation Biology courses that effectively integrate legal and policy aspects of conservation planning will count toward this category. Courses that are tools supporting professional practice, for example, Landsat, introductory GIS techniques, or more general courses such as environmental science, resource management, law enforcement, criminology, political science, and introductory survey courses in conservation will not apply.

Professional Experience

A minimum of five years of full-time professional-level wildlife experience, obtained within the last 10 years, is required. The professional wildlife experience must demonstrate the application of current biological knowledge to problems dealing directly with the wildlife resource—such as administration, education, research, or management—as a significant portion of job responsibilities. Technician-level work, such as data collection, surveys, and habitat manipulation conducted under existing protocols or under the specific direction of another, is not considered professional-level experience. Time spent obtaining advanced academic degrees can apply toward professional experience if the area of study is deemed relevant to the wildlife profession. A maximum of one year of professional experience will be granted for a master's degree, a maximum of two years for a PhD, and a maximum of three years for a master's and a PhD thesis or dissertations. Candidates must submit copies of abstracts or research summaries of thesis and/or dissertation(s) along with their application so the Certification Review Board may determine if the work is relevant to the wildlife profession. When time intervals for education and employment overlap, a detailed explanation must be provided. Professional experience credit can simultaneously be granted for a job and advanced degree provided the job is independent of the degree.

Partial credit may be granted for experience gained in positions peripheral to wildlife such as forester, range conservationist, soil conservationist, naturalist, environmental specialist, and consultant when a significant portion of the job responsibilities are those expected of a professional wildlife biologist. Experience credit will not be granted for positions such as high school biology teachers, park managers, fisheries biologists, or field or laboratory technicians. Experience credit also will not be granted for wetland delineation work unless it specifically addresses wildlife management.

Up to 12 months of volunteer experience will be credited toward the 5-year experience requirement provided that the position constitutes professional wildlife duties described in this section and is supported by a letter from the supervisor.

Professional Activity in Organized Activities and Mentorship

Recertification as a Certified Wildlife Biologist requires a minimum of 80 hours of participation in organized activities or mentorship. A minimum of 60 of these hours must be participating in organized activities including technical sessions at professional meetings, workshops, seminars, symposia, short courses, distance education, or college courses. The topics of these activities must be wildlife management, wildlife biology, or a closely related field. Up to 20 hours of mentorship may be applied to the 80 hours of required activities. Mentorship includes direct communication, or meeting, with a mentee discussing strategies for self-improvement or topics of technical or academic knowledge.

American Fisheries Society: Certified Fisheries Professional

Professional certification is limited to American Fisheries Society members. The following are the requirements for applicants who completed their B.S. /B.A. (or equivalent) degree on or after July 1, 2002. Additionally, a minimum of two years of appropriate work experience is required. Recertification requires active engagement in the fisheries profession and documentation of that engagement.

Education

Minimum coursework requirements completed with a "C–" or better grade with no pass/fail courses allowed.

- Fisheries and Aquatic Sciences: four courses, two of which must be directly related to fisheries sciences and at least one which must cover the principles of fisheries science and management.
- Other Biological Sciences: courses which when added to the above courses must total 30 semester hours.
- Physical Sciences courses must total 15 semester hours.
- Mathematics and Statistics courses must total 6 semester hours and include one calculus and one statistics, or two statistics courses.
- Communications courses must total 9 semester hours.
- Human Dimensions courses must total 6 semester hours

Professional Experience

For applicants applying with the required coursework, the experience requirement is five years for those with a bachelor's degree, four years for those with a master's degree, and two years for those with a doctorate.

Professional Development Activity

The American Fisheries Society uses Professional Development Quality Points (PDQPs) as the basis to gage professional development. A total of 30 PDQPs over the previous two years, or 100 PDQPs over the previous five years, are required. For the requirement of 30 PDQP over two years, a minimum of 10 of these points, and a maximum of 20, must fall under categories I & II; a minimum of 10 of these points, and a maximum of 20 points, must fall under categories III, IV, and V.

Certification Requirements

Category I: Continuing education—fisheries 0.5 PDQPs per hour of instruction This category includes subjects directly related to fisheries science or management, such as fisheries management, habitat management, fisheries economics, fish diseases, aquaculture or fish culture, fisheries policy and law, or aquatic ecology. Courses or training programs may be conducted by commercial or professional organizations, agencies, employers, or universities.

Category II: Continuing education—non-fisheries 0.5 PDQPs per hour of instruction This category includes subjects that are not primarily fisheries oriented but are professionally enriching to the individual, such as computer science and statistics, managerial and leadership skills, public speaking, problem solving, public relations, marketing, planning, and other related natural resource disciplines such as forestry or wildlife. Courses or training programs may be conducted by commercial or professional organizations, agencies, employers, or universities.

Category III: Oral communications in fisheries and non-fisheries subjects This category includes the development, preparation, and presentation of activities such as those described in categories I and II above for a meeting that is open to the public or a select group of invited participants. For fisheries subjects, the audience need not be fisheries professionals, but for non-fisheries subjects, the audience must be fisheries professionals. Points are earned at the following rates, author or coauthor of an oral or poster presentation at a professional meeting or to a non-professional audience yields 7 PDQPs. Organizer or instructor of a short course or workshop yields 20 PDQPs and instruction of a quarter, or semester, course yields 10 PDQPs per credit with a maximum of 30 PDQPs. Author or producer of self-instruction audiovisuals in fisheries yields 20 PDQPs.

Category IV: Written communications This category includes developing, writing, editing, reviewing, and publishing fisheries-oriented materials. The written material need not be published, but must be readily available to both professional and nonprofessional audiences. Points are earned at the following rates: Author or coauthor of a peer reviewed article or book chapter yields 15 PDQPs. Author or coauthor of a book or monograph 30 PDQPs. Editor or coeditor of a book or monograph 15 PDQPs. Author or coauthor of non-peer reviewed article in a magazine, brochure, newspaper 7 PDQPs. Author or coauthor of an agency publication or report 10 PDQPs. Reviewer or editor of an article that has been submitted for publication 3 PDQPs. Book reviewer for a professional publication 5 PDQPs.

Category V: Service This category involves membership and active participation in fisheries or aquatic professional societies and organizations, and community service that uses an individual's professional expertise in fisheries. Community service may include contributions of professional expertise to civic groups, environmental organizations, or government organizations. Points are earned for each year at the following rates: Serving in the highest office in an organization or subdivision of an organization; examples are president, director, chair, or journal editor yields 15 PDQPs. Serving in other offices in an organization such as secretary, treasurer, associate editor, newsletter editor, or committee chair yields 10 PDQPs. Committee membership yields 4 PDQPs. Serving as a mentor in the Hutton Junior Fisheries Biology Program yields 10 PDQPs.

Society for Range Management: Certified Professional in Rangeland Management

Certification as a Professional in Rangeland Management requires the appropriate education, experience, and passage of an examination. Recertification requires continuing education.

Education

At least one course should be taken in each of the following areas. If some courses meet more than one category of subject matter the credits may be apportioned to the appropriate categories.

- Rangeland Plant Identification: principles of plant taxonomy, and the use of keys and/or sight identification.
- Rangeland Vegetation Management: manipulating or establishing vegetation or habitat by means of grazing management, fire, chemical, or mechanical treatment.
- Rangeland Animal Management: controlling the intensity, timing, or distribution of animal use for animal production or other resource objectives.
- Rangeland Ecology: vegetation dynamics, plant-soil relationships, fire, herbivory, etc.
- Plant Physiology: basic processes, grazing and fire effects on plants, autecology.
- Rangeland or Natural Resource Planning/Policy: management plans, analysis of historic and current legal and policy effects on rangeland use.

- Rangeland Vegetation Measurement: vegetation measurement techniques, range condition/health assessment, monitoring and inventory.
- Soil Science: principles of soil science and soil classification.
- Range Economics or Microeconomics and Natural Resource/Environmental Economics: economic applicable to business or project-level analysis.
- Interpersonal Communication and Discussion: speech, technical writing, media techniques, conflict resolution, etc.
- Any other course which is considered as having special value for a rangeland professional; for example, watershed management, recreation, or forestry.

Professional Experience

Five years of appropriate work experience is required. A master's or doctorate degree in range management may be substituted for two year of work experience.

Continuing Education

The continuing education requirement is 32 Continuing Education Units during the two-year recertification period. At least half the credits should be directly related to the subjects described under the educational requirements above. The remainder of the credits should be in subject matter related to the professional practice of rangeland management. Units are awarded at the rate of one unit per hour participating in a short course, seminar, or workshop and one unit for every three hours in a field trip or tour. Preparation of a presentation is awarded units at a rate of three units per hour of presentation.

WORKS CITED

American Fisheries Society. n.d. Professional Certification Application. Available as a downloadable PDF at https://fisheries.org/membership/afs-certification/

Society for Range Management. n.d. Procedures for certification as a professional in rangeland management. Available as a downloadable PDF at http://rangelands.org/committees/cprm-committee/

Society of American Foresters. 2017a. Requirements. Accessed September 16, 2017. https://www.eforester.org/Main/Certification_Education/Certified_Forester/Requirements/Main/Certification/Requirements.aspx?hkey=7eae8378-e92b-438e-aba9-93e713cb38cc

Society of American Foresters. 2017b. Related Course Work. Accessed September 16, 2017. https://www.eforester.org/Main/Certification/Related_Course_Work.aspx

Society of American Foresters. 2017c. Experience Requirement. Accessed September 16, 2017. https://www.eforester.org/Main/Certification/Experience_Requirement.aspx

The Wildlife Society. 2017. Certification Programs Certified Wildlife Biologist Application. Accessed September 16, 2017. http://wildlife.org/wp-content/uploads/2017/08/CWB-Certification-August-2017-Restricted.pdf

Index

Academic readiness, 92
Academic requirements, 17, 20, 24, 29–30, 32
  degrees, 78–81
Accounting, 107
Accreditation, 15, 87
Adirondack Park, 72
Administration, administrators, 12, 26
Advisors, academic and career, 94
AFS. *See* American Fisheries Society
Agencies, 62
  history of, 115
  objectives, 12
  See also individual agencies
Agricultural Extension Services, 81
Alaska Native Corporations, 71
American Fisheries Society (AFS), 21, 48, 88, 91–92, 95
American Forestry Association, 48
American Forestry Congress, 48
American Institute of Fisheries Research Biologists, 21
American Ornithological Society, 29
American Society of Ichthyologists and Herpetologists, 29
American Society of Limnology and Oceanography, 21
American Society of Mammalogists, 29
Americans with Disabilities Act, 56
Animal and Plant Health Inspection Service (APHIS), 69
Animals, safety issues, 111
Animal sciences courses, 20
Animal Welfare Act, 69
Antiquities Act (1906), 50
Anza-Borrego State Park, 72

Application process, deadlines, 93
Applied sciences, 85
Area foresters, 14
Art movements, nature in, 47
Associate of Applied Science (AAS), 78
Associate of Arts (AA), 78–79
Associate of Science (AS), 78
Associate's degrees, 78–79
Association of Consulting Foresters, 15
Association of Field Ornithologists, 29
Association of Fish and Wildlife Agencies, 72
Audubon Society, 47

Baby boom, 53
Bachelor of Arts (BA), 79
Bachelor of Science (BS), 79
Bachelor's degrees, 79
  natural resources curriculum for, 82–87
Biltmore School, 52–53
Biology, 17
Bird preserves, 50
Bird watching, 3
Birge, Edward, 52
Bison, preservation of, 47
Bitterroot National Forest, 57
Boating safety, 111
Bonneville Power Administration (BPA), 71
Bookkeeping, 107
Boone and Crocket Club, 47, 49
Bureau of Biological Survey, 67
Bureau of Fisheries, 67

141

Bureau of Forestry, 50
Bureau of Indian Affairs (BIA), 58, 68, 71
Bureau of Land Management (BLM), 58, 66–67
Bureau of Reclamation (USBR), 68

Cal Fire, 72
California, 17, 59
California Department of Fish and Game/Wildlife, 72
California Department of Forestry and Fire Protection, 72
California Environmental Quality Act, 56
California State Parks, 72
Calling, versus career, 6
Career path, 93–94
Carson, Rachel, *Silent Spring*, 55
Catalogs, college and university, 81
Certification, certificates, 12, 15, 78, 87–88, 107
Chainsaws, 114
CIP. *See* Classification of Instructional Programs
Civil engineers, 14–15
Civilian Conservation Corps (CCC), 53
Classification of Instructional Programs (CIP), 82, 83 (table)
Clean Air Act, 55
Clean Water Act, 58
Clearcutting, 57
Clements, Fredric, 52
Cleveland, Grover, 49–50
Coal, as energy source, 46
Coalition of Natural Resource Societies, 90
  Natural Resource Education and Employment Conference, 88
Codes of ethics, 12
College requirements, 8–9, 12, 41
  fisheries managers, 24
  foresters, 17
  General Schedule, 40 (table)
  range managers, 20
  range technicians, 32
  wildlife managers, 29–30
Colleges, 73, 81
  academic terms, 80
  core courses, 85–86
  credits, 80
  majors, 78, 81–82
  programs, 77–80
Communication(s), 11, 12, 26
  skills in, 103, 104–105
Community colleges, 81
Compass, use of, 109
Compliance foresters, 14
Conduct, codes of, 105
Conservation, 46
  global concerns, 57–59
  management period, 53–56
  public role in, 47–49
  wise-use period, 49–53
Conservation movement, 8, 44–45, 50, 59–60
  Hetch Hetchy controversy, 51–52
Continuing education, 92
Cook County Forest Preserves, 73
Cornell University, 53
Cowles, Henry, 52
Crater Lake National Park, 50
Crews, wildland fire, 38–39
Critical thinking skills, 12
Culverts, blocked, 113
Curriculum, 81, 94
  advanced courses, 86–87
  bachelor's degrees, 82–85
  core courses, 85–86

Darwin, Charles, *Origin of Species*, 47
Decision making, context of, 6
Degrees, types of, 78–81
Department of Commerce, 70
Department of Defense (DOD), employment in, 69–70
Department of Energy (DOE), 70
Dingell-Johnson Act, 53
Doctor of Philosophy (PhD), 80
  driving, rural, 111–13
Duck Stamp program, 67

Earth Day, 57
Ecology, 8, 47, 52, 54, 82
Economists, wildlife, 26
Ecosystem(s), 54, 108
Education, 8–9, 12, 17, 20, 24, 29–30, 32, 36, 77–78, 98, 107
  for General Schedule positions, 40–41
  natural resources, 52–53, 88–92
  tips for success in, 92–95

Index

Eisenhower, Dwight D., 54
Elements of Forestry (Hough), 49
Elton, Charles, 52
Email etiquette, 105
Emerson, Ralph Waldo, 46
  Nature, 47
Employment, 8, 75, 107
  federal government, 64–71
  options for, 62–63
  position types, 39–40
  private, 73–74
  state and local government, 71–73
  statistics, 16, 20, 23, 29, 34, 35 (table), 63–64
Endangered Species Act, 57
Endangered Species Program, 67
Energy sources, 46
Engine crews, 38
Engines, maintenance and repair of, 113–14
Environmental Protection Agency (EPA), 71
Environmental Quality Improvement Act, 57
Equipment, maintenance and repair, 113–14
Ethics, codes of, 105
Etiquette, basic and professional, 105–106
Examinations, 17
Experience, 17, 36, 99, 104, 107
Extension services, 73

Federal government, 5, 11, 15, 103
  conservation, 47–48, 49
  employment by, 64–71
  employment levels, 63, 64
  job classification, 13–14, 104
  job website, 100
  Pathways Program, 102
  technical vs. professional work in, 31–32
  tours of duty and position types, 39–40
Federal Land Assistance, Management, and Enhancement Act (FLAME), 58
Fernow, Bernhard, 49, 53
Field and Stream (magazine), 47
Field labs, 94
Field skills, developing, 98–99

Firearms safety, 111
Firefighters, wildland, 36–39
First aid training, 109
Fish and Wildlife Administration, fisheries manager job categories, 21–22
Fish biology, 22
Fish Conservation and Management Act, 57
Fisheries Education in the 21st Century: Accommodating Change, 88, 91–92
Fisheries management, 1, 2, 6, 7, 8 (fig.), 10, 11
  description of, 21–24
  educational programs in, 90, 91–92
  employment in, 64 (table), 75
  programs in, 52, 53
Fishing, as social license, 6
FLAME. *See* Federal Land Assistance, Management, and Enhancement Act
Footwear, appropriate, 109–110
Forbes, Stephan Alfred, 52
Forensic specialists, wildlife, 25
Forest biology, courses in, 17
Forest engineers, 14–15
Forest Reserve Act, 49–50
Forest Reserve Organic Act, 50
Forest Reserves, 49–50
Forest resource measurements and inventory, 17
Forestry, educational programs in, 52–53, 89–90
Forestry Aid, seasonal firefighters, 36–37
Forestry Division, 49
Forestry extension agents, 14
Forestry profession, foresters, 1, 2, 7, 6, 7, 10, 11, 14
  description of, 15–18
  employment of, 63 (table), 75
  occupations of, 14–15
Forestry technician, 11, 33–34, 37
Forests, definition of, 4
Forest Stewards Guild, 15
4H Programs, 81
Four-wheel drive, familiarity with, 113
Froese, Robert, 6
Fuels crews, 38
Full-time firefighters, 37

Fundamentals of Ecology (Odum), 54
Fur-bearing animals, trapping, 6
Future of the Wildlife Profession, The, 91

Game Management (Leopold), 54
Game species, regulation of, 5
General Land Office, 66
General Natural Resources Management and Biological Sciences, 30
General Schedule (GS) positions, 32, 40–41, 104
Geographic Information Systems (GISs), 26, 118
GI Bill, 54
Global concerns, 57–59
Governor's Conference on the Conservation of Natural Resources, 50
GPS, proficiency with, 109
Grand Canyon National Monument/ Park, 50
Grasslands, 4
Grazing, 3, 6
Grazing Service, 66
Great Depression, 53
Great Lakes region, 4
Grey Towers, 53
Grinnell, George Bird, 47
Grooming, 106
Group Standards, 11

Haeckel, Ernst, 47
Hand crews, 38
Hand tools, safety and maintenance of, 114
Harrison, Benjamin, 49
Hazards, overhead, 111
Healthy Forest Restoration Act, 58
Heavy equipment, safety around, 114
Helitack crews, 38
Herpetologists League, 29
Hetch Hetchy Valley, 51–52
Hotshot crews, 38
Hough, Franklin, *Elements of Forestry*, 49
How-To-Guide for Pursuing a Career in Natural Resources, A, 99–100
  tips from, 102–104
Hudson River School, 47
Hunting, 6, 111
  market, 49

Hunting and fishing clubs, 47, 48, 49
Hutton Junior Fisheries Biology Program, 99

Indiana Dunes, 52
Indian Forestry & Natural Resources National Directory, 71
Industrialization, 46
Insects, safety around, 111
Inspectors, wildlife, 25
Interagency hotshot crews, 38
Interdisciplinary sciences, 85
International Whaling Commission, 54
Internships, 99, 103
Interpretation, park ranger, 36
Interstate Highway System, 54
Intertribal Timber Council, 71
Interviewing, job, 100–101
Inventory foresters, 15

Jobs, 30
  applying for, 99–100
  education requirements and salaries, 40–41
  federal definitions, 39–40
  fisheries, 21–23
  forestry, 14–16
  interviewing for, 100–101
  range management, 18–20
  references, 101–102
  related professional and technical, 31–36
  requirements and skills, 13–14
  wildland firefighting, 36–39
  wildlife management, 25–28
Junior college, 81

Kyoto Protocol, 59

Lacy Act, 52
Land
  federal protection of, 47–48, 49
  and resource management, 3–4
  state and local, 72
Land and Water Conservation Act, 55
Land ethic, 54
Land-grant universities, 73, 81
Landscape artists, 47
Land-use plans, 15–16, 20
Law enforcement, wildlife, 25

Index

Lead forestry technician, 33
Leopold, Aldo
  Game Management, 54
  Sand County Almanac, 54
Licensing, 12
  forester, 17–18
Lincoln, Abraham, Yosemite Grant, 47
Local governments, employment, 63, 64, 72–73

Management, 36
Management period, 53–54
Man and Nature (Marsh), 47
Marine Mammal Protection Act, 57
Market hunting, 49
Marsh, George Perkins, *Man and Nature*, 47
Master of Arts (MA), 79
Master of Forestry (MF), 80
Master of Science (MS), 79–80
Master's degrees, 79–80
Mathematics, 85
McIntire-Stennis Cooperation Research Act, 55
McKinley, William, 50
Measurement
  of resources, 11
  units of, 114
Mesa Verde National Park, 50
Michigan Department of Natural Resources, 72
Migratory Bird Conservation Act, 52
Migratory Bird Hunting Stamp Act, 53
Migratory Bird Management program, 67
Migratory Bird Treaty Act, 52
Mining, 3
Monongahela National Forest, 57
Montreal Protocols, 59
Mt. Rainier National Park, 50
Muir, John, 48–49, 51
Multiple-Use, Sustained-Yield Act, 55

Name Change Act, 50
National Association of University Fisheries and Wildlife Programs (NAUFWP), 90, 91
National Association of University Forest Resources Programs (NAUFRP), 89
National Environmental Policy Act, 56

National Fish Hatchery System, 67
National Forest Commission, 50
National Forest Management Act, 57
National Forests, 50
National Historic Preservation Act, 55
National Indian Forest Resource Management Act, 58
National Landscape Conservation System, 58
National monuments, 50
National Oceanographic and Atmospheric Administration (NOAA), 66, 70
National Outdoor Recreation Act, 55
National parks, 47, 48, 49, 50
National Parks Omnibus Management Act, 58
National Park Service (NPS), 5, 51, 52, 55, 66, 67–68
National Trails Act, 55
National Wildlife Refuge System, 67
Natural Resource Conservation Service (NRCS), 68–69
Natural resource scientist, 11, 30
Natural resources, 1, 2, 45
  defining, 3–4
  education in, 52–53
  policy, 8
  US use of, 4–5
Natural resources programs
  accreditation and certification, 87–88
  changes in, 88–89, 90–92
  curriculum, 82–87
  enrollment in, 59, 89–90
  tips for success in, 92–95
Natural Resources Management and Biological Sciences Group (400 job group), 13–14
Natural sciences, 47, 85
Nature, appreciation for, 47–48
Nature (Emerson), 47
Nature Conservancy, 73
Navigation, 109
Networking, 94, 102–103
New England, timberland in, 4
New York State, 72
NOAA. *See* National Oceanographic and Atmospheric Administration
Nonextractive resources, 3
Nongovernment organizations, 73
Nonrenewable resources, 3

North American Summit on Forest Science Education, 88, 90
Northwest Forest Plan, 57
Notetaking, 106–7
NRCS. *See* Natural Resource Conservation Service

Odum, Eugene, *Fundamentals of Ecology*, 54
Office of Environmental Quality, 57
Office of Personnel Management (OPM), job classification, 13–14
Office of Personnel Management List of Occupational Standards, 11
Operations foresters, 15
OPM. *See* Office of Personnel Management
Oregon, 52, 72
Origin of Species (Darwin), 47
Outreach specialists, wildlife, 25
Overhead hazards, 111

Pacific Northwest, 4
Pacing, 109
Paris Climate Agreement, 59
Park ranger, as profession, 11, 34–36
Parks, defined, 35
Part-time positions, 103
Pathways Program, 102
Pay rates, federal, 11, 40–41
Permanent career-seasonal positions, 39
Permanent full-time positions, 39
Permanent part-time positions, 39
Persons with disabilities, federal employment, 104
Petroleum, as energy source, 46
Physics, 85
Pinchot, Gifford, 50–51
Pittman-Robertson Act, 53
Plant sciences courses, 20
Policy decisions, 45
Prerequisites, 94
Prescribed burns, as social license, 6
Prescribed wildland fire crews, 39
Private industry, 103, 105
Private land, 4
Professional associations, 12, 18, 21, 28–29
Professional certification, 87–88
Professional degree, 80
Professional full-time firefighters, 37
Professional scientific work, 12
Professional Work in the Natural Resources Management and Biological Sciences Group, 11
Professions, professional work, 2, 6–7, 8, 10–11, 31–32, 34–36
 job requirements of, 13–14
 See also by profession
Public lands, 4, 47–48, 49, 66, 72
Public relations specialists, wildlife, 26
Public trust doctrine, 5
Pure sciences, 85

Quality of life, 55–56
Quasi-government agencies, employment, 71

Radio use, 114
Range, as renewable, 3
Rangeland, 4
 management of, 18–20
Range management, 1, 2, 6, 7–8, 10, 11, 90
 description of, 18–20
Range management courses, 20
Range technician, 11, 32
Real Estate Investment Trust (REIT), 74
Records, keeping, 106–7
Recreation, 3, 5, 54
Recreation activity day, 5
Recreation foresters, 15
Reference materials, maintaining, 108
References, 101–102
Refuge managers, wildlife, 27–28
Registered Professional Forester (RPF), 17
REIT. *See* Real Estate Investment Trust
Relationships, building, 102–103
Renewable resource management, courses in, 17
Renewable resources, 3
Resource management, 11, 88–89
Resource management studies, courses in, 20
Resources, measuring, 11
Responsibilities
 fisheries managers, 21–23
 foresters, 15–16
 range managers, 18–20

Résumé, 100, 103
Roosevelt, Franklin, 53
Roosevelt, Theodore, 48 (fig.), 49, 50, 51 (fig.)
RPF. *See* Registered Professional Forester
Rural areas, driving in, 111–13

SAF. *See* Society of American Foresters
Safety, outdoor, 109–15
Salaries, 16, 40–41
Sand County Almanac (Leopold), 54
San Francisco, and Hetch Hetchy Valley, 51
Schenk, Carl, 52
Science(s), 12, 82, 85
Scientists, 57
Seasonal firefighters, 36–37
Service foresters, 14
Sierra Club, 49
Silent Spring (Carson), 55
Silvicultural foresters, silviculturists, 15
Siuslaw National Forest, 69 (fig.)
Skills, 100
  communication and analytical, 103
  field, 98–99, 108–115
  human dimension, 104–108
  social, 12
Smokejumpers, 38
Social context, 6
Social license, 6
Social sciences, 82, 85
Society for Marine Mammalogy, 29
Society for Range Management (SRM), 18, 54, 87, 88
Society for the Study of Amphibians and Reptiles, 29
Society of American Foresters (SAF), 15, 51, 87, 88
Soil sciences courses, 20
SRM. *See* Society for Range Management
State governments, employment by, 71–72
States, forester licensing requirements, 17–18
Students, tips for, 92–95
Supervisory forestry technician, 33–34
Survival, field, 114

Taft, William Howard, 51
Tansley, Arthur, 54
Teaching, 62
Teamwork, 106
Technical jobs/work, technicians, 31, 32–34
Temporary/seasonal positions, 40
Tennessee Valley Authority (TVA), 71
Terms of service, 11
Thoreau, Henry David, 46, 52
  Walden, 47
Timber harvesting, 6, 46, 54
Timber Investment Management Organizations (TIMOs), 74
Timberland, 4, 74
Tools, safety and maintenance of, 114
Tours of duty, federal jobs, 39–40
Traditional natural resources profession, 11
Training, 12, 32, 36, 38, 107, 109
Transcendental movement, 46–47
Transfer Act (1905), 50
Transmissions, manual, 113
Trapping, social license for, 6
Tribal governments, 71
TVA. *See* Tennessee Valley Authority

United Nations Framework Convention on Climate Change, 59
Universities, 59, 73, 81
  natural resource education, 52–53
  programs, 77–80
University of California, Berkeley, North American Summit on Forest Science Education, 88, 90
University of Washington, fisheries program at, 53
Urban forests, 4, 5
US Army Corps of Engineers (USACE), employment, 69–70
US Commission of Fish and Fisheries, 49
US Department of Agriculture (USDA), 49, 105
  employment in, 68–69
  Gifford Pinchot and, 50–51
US Department of the Army, 35, 48
US Department of the Interior, 35, 38, 50
  employment in, 66–68
  and Yellowstone National Park, 47–48

US Fish and Wildlife Service (USFWS), 66, 67, 72
US Forest Service (USFS), 3, 4, 50, 53, 55
  employment by, 64, 66, 68
US Geological Survey (USGS), 68
USACE. *See* US Army Corps of Engineers
USAJOBS, 100
USBR. *See* Bureau of Reclamation
USFS. *See* US Forest Service
USFWS. *See* US Fish and Wildlife Service
Utility foresters, 14

Values, economic vs. social, 85
Vehicles, fieldwork and, 113–14
Vests, cruising, 110
Veterans, and federal employment, 104
Vienna Convention for the Protection of the Ozone Layer, 59
Visitor protection and service, park ranger, 36
Visual arts, 47
Volunteering, 99, 103, 107

Walden or Life in the Woods (Thoreau), 47
Washington, 59
Water, 3, 4, 5
Waterfowl, 5
Water Quality Act, 55
Water safety, 111
Weather safety, 111
Weeks Act, 52
Weyerhaeuser, 74, 105
Wild and Scenic Rivers Act, 86
Wilderness Act, 55
Wilderness Areas, 56, 58
Wild fires, suppression of, 6
Wildland firefighting, 11, 36–37, 110
  crew types, 38–39
Wildland Fire Jobs website, 38
Wildland fire modules, 39

Wildlands, 3–4
Wildlife, 3, 4–5
Wildlife biologist, 25
Wildlife biology, 7
Wildlife management, 1, 2, 5, 6, 7, 10, 11, 24, 54
  description of, 26–30
  educational programs in, 90, 91
  employment, 23, 64 (table), 75
  occupations in, 25–26
Wildlife policy analyst, 26
Wildlife preserves, 50
Wildlife public educator and outreach specialist, 25
Wildlife Refuge Management, job descriptions, 27–28
Wildlife Society, 28–29, 54, 88, 91
Wildlife technician, 25
Wisconsin, county forests in, 73
Wise-use period
  federal forest preservation and, 49–50
  and Hetch Hetchy Valley, 51–52
  and natural resource education, 52–53
  Gifford Pinchot and, 50–51
Wood scientists, 15
Works Progress Administration (WPA), 53
World War II, 53
WPA. *See* Works Progress Administration
Writing, skills in, 104, 105
Wyoming Game and Fish Department, 72

Yale School of Forestry, 51, 53
Yellowstone National Park, 48
Yellowstone National Park Act, 47
Yosemite Grant, 47
Yosemite National Park, 47, 48, 49
  Hetch Hetchy valley and, 51–52
Youth Conservation Corps, 99

Zoologists, 23, 64 (table), 75